New Methods and Paradigms for Modeling Dynamic Processes Based on Cellular Automata

Stepan Mykolayovych Bilan
State University of Infrastructure and Technology, Ukraine

Mykola Mykolayovych Bilan
Mayakskaya Secondary School, Moldova

Ruslan Leonidovich Motornyuk
Main Information and Computing Center, Ukraine

A volume in the Advances
in Computer and Electrical
Engineering (ACEE) Book Series

Published in the United States of America by
    IGI Global
    Engineering Science Reference (an imprint of IGI Global)
    701 E. Chocolate Avenue
    Hershey PA, USA 17033
    Tel: 717-533-8845
    Fax: 717-533-8661
    E-mail: cust@igi-global.com
    Web site: http://www.igi-global.com

Library of Congress Cataloging-in-Publication Data

Names: Bilan, Stepan Mykolayovych, 1962- author. | Bilan, Mykola
  Mykolayovych, 1961- author. | Motornyuk, Ruslan Leonidovich, 1976-
  author.
Title: New methods and paradigms for modeling dynamic processes based on
  cellular automata / by Stepan Mykolayovych Bilan, Mykola Mykolayovych
  Bilan, and Ruslan Leonidovich Motornyuk.
Description: Hershey, PA : Engineering Science Reference, [2020] | Includes
  bibliographical references and index. | Summary: "This book describes
  the theoretical concepts of constructing asynchronous cellular automata
  with active cells. It also examines the use of theoretical principles
  for solving modeling problems and solving specific applied problems of
  forming pseudorandom sequences and image processing based on modeling of
  the human visual channel" -- Provided by publisher.
Identifiers: LCCN 2019042065 (print) | LCCN 2019042066 (ebook) | ISBN
  9781799826491 (hardcover) | ISBN 9781799826507 (paperback) | ISBN
  9781799826514 (ebook)
Subjects: MESH: Computer Simulation | Image Processing, Computer-Assisted |
  Pattern Recognition, Automated | Electronic Data Processing
Classification: LCC QA 267.5.C45 (print) | LCC QA 267.5.C45 (ebook) | NLM
  QA 267.5.C45 | DDC 511.3--dc23
LC record available at https://lccn.loc.gov/2019042065
LC ebook record available at https://lccn.loc.gov/2019042066

This book is published in the IGI Global book series Advances in Computer and Electrical Engineering (ACEE) (ISSN: 2327-039X; eISSN: 2327-0403)

British Cataloguing in Publication Data
A Cataloguing in Publication record for this book is available from the British Library.

All work contributed to this book is new, previously-unpublished material.
The views expressed in this book are those of the authors, but not necessarily of the publisher.

For electronic access to this publication, please contact: eresources@igi-global.com.

Advances in Computer and Electrical Engineering (ACEE) Book Series

ISSN:2327-039X
EISSN:2327-0403

Editor-in-Chief: Srikanta Patnaik, SOA University, India

MISSION

The fields of computer engineering and electrical engineering encompass a broad range of interdisciplinary topics allowing for expansive research developments across multiple fields. Research in these areas continues to develop and become increasingly important as computer and electrical systems have become an integral part of everyday life.

The **Advances in Computer and Electrical Engineering (ACEE) Book Series** aims to publish research on diverse topics pertaining to computer engineering and electrical engineering. **ACEE** encourages scholarly discourse on the latest applications, tools, and methodologies being implemented in the field for the design and development of computer and electrical systems.

COVERAGE

- Circuit Analysis
- VLSI Fabrication
- Algorithms
- Applied Electromagnetics
- Chip Design
- Analog Electronics
- Digital Electronics
- Sensor Technologies
- Computer Hardware
- Microprocessor Design

IGI Global is currently accepting manuscripts for publication within this series. To submit a proposal for a volume in this series, please contact our Acquisition Editors at Acquisitions@igi-global.com or visit: http://www.igi-global.com/publish/.

The Advances in Computer and Electrical Engineering (ACEE) Book Series (ISSN 2327-039X) is published by IGI Global, 701 E. Chocolate Avenue, Hershey, PA 17033-1240, USA, www.igi-global.com. This series is composed of titles available for purchase individually; each title is edited to be contextually exclusive from any other title within the series. For pricing and ordering information please visit http://www.igi-global.com/book-series/advances-computer-electrical-engineering/73675. Postmaster: Send all address changes to above address. Copyright © 2021 IGI Global. All rights, including translation in other languages reserved by the publisher. No part of this series may be reproduced or used in any form or by any means – graphics, electronic, or mechanical, including photocopying, recording, taping, or information and retrieval systems – without written permission from the publisher, except for non commercial, educational use, including classroom teaching purposes. The views expressed in this series are those of the authors, but not necessarily of IGI Global.

Titles in this Series

For a list of additional titles in this series, please visit:
http://www.igi-global.com/book-series/advances-computer-electrical-engineering/73675

Design and Investment of High Voltage NanoDielectrics
Ahmed Thabet Mohamed (Aswan University, Egypt & Qassim University, Saudi Arabia)
Engineering Science Reference • © 2021 • 340pp • H/C (ISBN: 9781799838296) • US $195.00

Research Advancements in Smart Technology, Optimization, and Renewable Energy
Pandian Vasant (University of Technology Petronas, Malaysia) Gerhard Weber (Poznan University of Technology, Poland) and Wonsiri Punurai (Mahidol University, Thailand)
Engineering Science Reference • © 2021 • 407pp • H/C (ISBN: 9781799839705) • US $225.00

HCI Solutions for Achieving Sustainable Development Goals
Fariza Hanis Abdul Razak (Universiti Teknologi MARA (UiTM), Malaysia) Masitah Ghazali (Universiti Teknologi Malaysia (UTM), Malaysia) Murni Mahmud (International Islamic University Malaysia, Malaysia) Chui Yin Wong (Multimedia University, Malaysia) and Muhammad Haziq Lim Abdullah (Universiti Teknikal Malaysia, Melaka, Malaysia)
Engineering Science Reference • © 2021 • 300pp • H/C (ISBN: 9781799849360) • US $195.00

Innovations in the Industrial Internet of Things (IIoT) and Smart Factory
Sam Goundar (The University of the South Pacific, Fiji) J. Avanija (Sree Vidyanikethan Engineering College, India) Gurram Sunitha (Sree Vidyanikethan Engineering College, India) K Reddy Madhavi (Sree Vidyanikethan Engineering College, India) and S. Bharath Bhushan (Sree Vidyanikethan Engineering College, India)
Engineering Science Reference • © 2020 • 300pp • H/C (ISBN: 9781799833758) • US $225.00

For an entire list of titles in this series, please visit:
http://www.igi-global.com/book-series/advances-computer-electrical-engineering/73675

701 East Chocolate Avenue, Hershey, PA 17033, USA
Tel: 717-533-8845 x100 • Fax: 717-533-8661
E-Mail: cust@igi-global.com • www.igi-global.com

Table of Contents

Preface

Cellular Automata ... Cellular ... Automata ...
...
What are they and what are they used for?
Who invented them and why?
How did you manage to implement them?
Isn't it easier to somehow do without them?..
Hmm, is it edible?
...

Such or very similar questions can often arise in a person who first heard the new term "cellular automaton". Some of these questions are both simple and complex at the same time, some do not have an unambiguously correct answer, some will be answered only in the future, but some already have answers, proven by theoretical principles and supported by practical implementation. And if You are reading this book, then once upon a time (or maybe just a few moments before) they also asked similar questions, but you want to find answers here. We, as the authors of this book, hope that the material presented below will allow the Reader to find the necessary answers, if not completely, then at least partially; professionals and experts in this field will be able to learn something new for themselves, and beginners will be interested in further studying the theory of cellular automata.

In modern society, we increasingly hear the phrase "Artificial Intelligence". However, in most cases, people do not think about the deep sense of its meaning.

The accelerating development of computer technology and communications can replace many of the functions of human intellectual activity, as well as help him in making decisions in various situations of his life. To implement intelligent functions for various purposes, various models, paradigms, architectures, and hardware and software are being developed. And since the world does not stand still and is constantly evolving, there is a need to constantly study various dynamic processes to determine possible negative situations that can lead to undesirable catastrophic phenomena and changes. To date, all dynamic processes are investigated using simulation. Especially those that require large resources to accurately reproduce them.

Recently, more and more attention has been paid to the study of natural processes in nature. More and more scientific works appear that describe the behavior and development of living organisms, as well as the processes of their interaction. Cellular automata are increasingly used to model them.

One of the most popular approaches to a limited description of the behavior of the universe is the "life" game proposed by Conway. In this game, using the cellular automaton, the formation of colonies of "living" cells in the universe is described. Moreover, all "living" cells have the same state. The development and disappearance of colonies depends on the initial state of the universe (cellular automaton). In this case, at least three "living" cells must be present in the initial state. Otherwise, no development of the universe happens. This model of life is of great scientific importance for modeling dynamic processes. However, due to its limitations it does not give a valid description of the processes of evolution of the universe.

This book is aimed at describing the models and paradigms of building cellular automata that allow you to simulate the dynamics of the interaction of living organisms from a different scientific point of view. For this, asynchronous cellular automata with a dynamically changing number of "living" cells are used. There are no restrictions in the initial state of the cellular automaton. The book describes such cellular automata.

The book has nine chapters. The chapters describe the theoretical concepts of constructing asynchronous cellular automata with active cells. Much attention is paid to the use of the proposed theoretical principles for solving modeling problems and solving specific applied problems of forming pseudorandom sequences and image processing based on modeling of the human visual channel.

Chapter 1 describes the basic theoretical principles for the theory of cellular automata. The history of the emergence of cellular automata based on an analysis of existing information sources is presented. The modern classification of cellular automata is presented.

Chapter 2 describes the approach to the modeling of dynamic systems based on cellular automata. The rules of the transfer of properties and conditions from cell to cell in cellular automata of various organizations are considered. The basic cell structures are presented during the transfer of not only states, but also the properties of cell activity and states. The options are considered when the cell itself selects a cell among the cells in the neighborhood that will become active in the next time step.

Chapter 3 describes the basic models and paradigms for constructing asynchronous cellular automata with one active cell. The rules of performing local state functions and local transition functions are considered. The basic cell structures during the transmission of active signals for various local transmission functions are presented.

In Chapter 4, an existing approach to modeling dynamic systems is described. The rules of transferring properties and conditions from cell to cell in cellular automata of various organizations are considered. The basic cell structures during the transmission of not only states, but also the properties of cell activity and states.

Chapter 5 focuses on analyzing a functioning model of an asynchronous cellular automaton with a variable number of active cells. The rules for the formation of active cells with new active states are considered. Codes of active states for the von Neumann neighborhood are presented, and a technique for coding active states for other forms of neighborhoods is described. Several modes of operation of asynchronous cellular automata from the point of view of the influence of active cells are considered.

Chapter 6 describes well known models and implementation options for pseudorandom number generators based on cellular automata. Pseudorandom number generators based on synchronous and asynchronous cellular automata are briefly reviewed. Pseudorandom number generators based on one-dimensional and two-dimensional cellular automata, as well as using hybrid cellular automata, are described. New structures of pseudorandom number generators based on asynchronous cellular automata with a variable number of active cells are proposed. Testing of the proposed generators was carried out, which showed the high quality of the generators.

Chapter 7 describes the use of asynchronous cellular automata with controlled movement of active cells for image processing and recognition. A time-pulsed image description method is described. Various models and structures of cellular automata for transmitting active signals are presented. Methods for describing images of individual plane figures, as well as a method for describing images consisting of many separate geometric objects, are proposed. The cellular automaton is considered as an analogue of the retina of the human visual canal. The circuitry structures of cells of such asynchronous cellular automata are presented, and the software implementation of the proposed methods is also performed.

Chapter 8 is devoted to the presented software that implements models of asynchronous cellular automata with a variable set of active cells. The software is considering one of the modifications of the Conway game of "Life". In the proposed model "New Life" the possibility of functioning of a separate "living" cell is realized, which, when meeting with other "living" cells, participates in the "birth" of new "living" cells with a different active state.

Chapter 9 describes the principles of functioning of asynchronous cellular automata with many active cells that are organized in colonies. The rules for the interaction of cell colonies with different and identical active states (the union of colonies, destruction of colonies and the formation of colonies) are presented. An example of the use of such asynchronous cellular automata for the description of contour images is described.

Stepan Mykolayovych Bilan
State University of Infrastructure and Technology, Ukraine

Mykola Mykolayovych Bilan
Mayakskaya Secondary School, Moldova

Ruslan Leonidovich Motornyuk
Main Information and Computing Center, Ukraine

Chapter 1
About Cellular Automata

ABSTRACT

The chapter describes the basic theoretical principles for the theory of cellular automata. The history of the emergence of cellular automata based on an analysis of existing information sources is presented. The modern classification of cellular automata is presented. The structures of elementary and two-dimensional cellular automata are described. In terms of the rules for the functioning of cellular automata, synchronous, asynchronous, and probabilistic cellular automata are briefly described. Researchers are presented who have made a significant contribution to the development of the theory of cellular automata.

INTRODUCTION

The chapter describes the basic theoretical principles for the theory of cellular automata. The history of the emergence of cellular automata based on an analysis of existing information sources is presented. The modern classification of cellular automata is presented. The structures of elementary and two-dimensional cellular automata are described. In terms of the rules for the functioning of cellular automata, synchronous, asynchronous and probabilistic cellular automata are briefly described. Researchers are presented who have made a significant contribution to the development of the theory of cellular automata.

DOI: 10.4018/978-1-7998-2649-1.ch001

THE HISTORY OF THE EMERGENCE THE THEORY OF CELLULAR AUTOMATA

In our world, any result that contributes to the rapid development of scientific and technological progress is based on the great work of many researchers who have invested a lot of work and in many cases have devoted their whole lives to moving forward in research. Therefore, it is important to know what preceded the formation of a particular theory, what ideas or scientific breakthroughs made it possible to obtain an existing result.

Today, a large number of scientists use cellular automata (CA) in their research. However, not everyone knows the history of the theory of CA. This chapter describes a brief history of the formation of the theory of CA.

So, let's start with a simple one - with the meaning of the term itself. The following definition of cellular automata is given in the book (Toffoli, Margolus 1991): «Cellular automata are discrete dynamical systems whose behavior is completely determined in terms of local dependencies». To a large extent, this is also the case for a large class of continuous dynamical systems defined by partial differential equations. In this sense, cellular automata in computer science are analogous to the physical concept of "field" ... a cellular automaton can be thought of as a stylized world. A uniform lattice, each cell, represents the space or cell of which contains several bits of data. Time goes forward in discrete steps, and the laws of the world are expressed by a single set of rules, say, a small look-up table according to which any cell at each step calculates its new state from the states of its close neighbors. Thus, the laws of the system are local and the same everywhere. "Local" means that in order to find out what will happen here a moment later, it is enough to look at the state of the immediate environment: no long-range action is allowed. "Identity" means that the laws are the same everywhere: "I can distinguish one place from another only in the form of the landscape, and not in any difference in the laws".

There is and not such a "dry" and more figurative description: "Cellular automata are stylized, synthetic worlds, defined by simple rules, similar to the rules of a board game". They have their own kind of matter, which is spinning in their own space and time. One can imagine the amazing diversity of these worlds. You can really build them and observe how they develop. Since we are inexperienced creators, it is unlikely that we will be able to get an interesting world on the first try; as people, we can have different ideas about what makes the world interesting, or what we might want to do with it. In any case, after we are shown the world of the cellular automaton, we will want to create it ourselves. By creating one, we will want to try to create another. After creating a few, we can create a world specifically designed for a specific purpose, with some confidence.

The machine of cellular automata is a synthesizer of worlds. Like an organ, it has keys and registers with which the instrument's capabilities can be powered, combined and rearranged, and its color screen is a window through which you can observe the world that is being "played".

Now, for a better understanding of the material, we turn to the background of the emergence of such a difficult definition.

It is generally accepted that the founder of the concept of cellular automata was John von Neumann. But a more detailed study of the historical prerequisites for the emergence of this topic allows us to conclude that the very idea of using this architecture appeared almost simultaneously (in the 1940s) at once by several scientists working on different tasks.

At the Los Alamos National Laboratory, John von Neumann worked on self-replicating systems. At the same time, in the same laboratory, another of her collaborators, Stanislav Ulam, was developing a mathematical model of the theory of crystal growth. It is not surprising that when these two scientists met, there was communication and mutual exchange of ideas.

In his early research on self-reproducing automatic machines, von Neumann described a machine that could live in a large soup with spare parts floating around it, and studied how such a machine could assemble its doubles from these parts. This type of automaton became known as a kinematic automaton and could be a prototype of what we usually call a robot. After discussions with his friend Stanislav Ulam, who studied crystal growth, Neumann decided to investigate instead a much simpler model of automata called cellular. The result of the discussion of the topic was the idea that a cellular automaton consists of an unlimitedly iterated, or mosaic, array of finite automata, each of which interacts with its neighbors. It represents a "space" in which events related to the functioning of the automaton take place, and for which we can formulate simple rules. From this a variety of cellular automata followed: the geometry of the cell space, neighborhood relations, types of automata, transition laws for the system, and the initial state of the system were defined differently. The main conclusion from the work of von Neumann and Ulam follows that on a system of cells with 29 states one can simulate universal constructing machines, universal computing machines and self-reproducing machines.

The third and fourth "parents" are considered approximately at the same time worked at the Massachusetts Institute of Technology (MIT) Norbert Wiener (Norbert Wiener, 1894–1964) and Arturo Rosenblut (Arturo Rosenblueth, 1900–1970) developed a cellular automaton model of an excitable medium to describe the propagation of impulses in nerve nodes (Wiener, Rosenbluth 1946).

The fifth "co-author" of the theory of cellular automata is rightfully considered an outstanding German engineer Konrad Zuse (Konrad Zuse, 1910–1995) – creator of the first in the modern sense of the programmable computer Z3 (1941) and the first high-level programming language (1945). He called cellular automata "computational spaces" and considered them as a possible architecture of computing systems. In 1969, K. Zuse published the book "Computable Cosmos" (Zuse, 1969), where he suggested that, by its nature, the Universe is a gigantic cellular automaton, and the physical processes occurring in it are the essence of the calculations. At that time, this view of the Universe was shocking, while now the idea of a self-calculating Universe was further developed (Berthold, 2009; Flake, 1998; Gernert 1989; Llachinski 2002; Kauffman 1995; Petrov P. 2003; Schiff 2008).

But the first fundamental work on the theory of cellular automata is still a book that was created according to rough records and incomplete articles by J. von Neumann, completed and revised by his long-term collaborator A. Burks AW, and was published in 1966 (Neumann, 1966).

Cellular automata, by virtue of their discreteness, are relatively easily modeled using computers and, as a result, in the 1950s to 1970s of a XX century gain popularity. Researchers of various scientific fields study and use cellular automata with various properties for various purposes. At this time, the main works were published, which laid the basis for the general theory of cellular automata. In 1970, Burks (Burks, 1970) in his book Essays on Cellular Automata set out the main terms and provisions on this topic, summarizing the multiple theoretical studies for that period. Subsequently, the theory was comprehensively studied and expanded by Aladev in 1974 (Aladyev, V. 1974), Smith (Smith 1976), Vollmar (Vollmar, T. (1977). These and other works set forth the "classical" theory of cellular automata — general propositions, theorems, and laws that are true for all areas of research at that time.

Since the 1980s, the study of cellular automata has taken on a more specialized connotation. On the basis of the general theory, various configurations of cellular automata for specific research areas are created and studied. Thanks to comprehensive research, it was possible to create a powerful mathematical theory aimed at classifying and studying the properties of various models. In 1983, S. Wolfram published his first work (Wolfram, 1983) on the statistical parameters of cellular automata. S. Wolfram further developed the mathematical aspects of this theory (Wolfram, 1986), which allowed him and other researchers to classify and study almost the entire set of cellular automata in detail.

It should be noted that cellular automata are not just machines working with a field divided into cells. The scope of cellular automata is almost limitless: from the simplest "tic-tac-toe" to artificial intelligence. The topic of cellular automata is very relevant, as it can lead to clues to many issues in the world.

Recently, cellular automata have become quite widely used to model various physical phenomena. The latter are systems consisting of identical cells. Each cell can only be in a finite set of states. Its evolution, i.e., the transition from one state to another, occurs at discrete moments of time and depends only on its own state at a given moment in time and on the state of its closest neighbors. The rules describing the evolution of cells are unusually simple. They are the same for all cells, regardless of their location, and do not change over time.

It turned out that these simplest mathematical models are capable of demonstrating behavior resembling processes occurring in nature in complexity: growth, reproduction, self-organization, diffusion, chemical interaction, flow of fluids (Wolfram, 1986a; Frish, et al 1987), the formation of fractal structures, etc.

There is a circle of tasks whose solution by traditional methods is difficult or impossible at all. These can be problems with complex geometry (in which problems arise with the specification of boundary conditions, or with a description of the field itself); tasks in which the process is non-linear and has a threshold character; tasks in which the description of variables in the form of continuous quantities is difficult due to the small number of their values.

The use of cellular automata to solve such problems is more appropriate. Typical areas of their use can be tasks associated with the description of chemical reactions with a large number of reagents (different reagents can be represented by particles of different types in cellular automata); reaction-diffusion systems; flow of immiscible liquids; filtration processes (a porous medium can be easily displayed in cellular automata); crystal growth modeling; modeling of processes of growth, reproduction and self-organization in biological systems: the study of self-organized criticality in distributed systems arising in biology, economics, and geology (studying the evolution of the genome, stock market, and the earth's crust "on the brink of chaos"); simulation of interacting spin systems.

5

Cellular automata are applicable not only in mathematics and physics, but also in biology (Csahyk, & Vicsek 1995), economics, sociology, computer science, etc. Using cellular automata, problems of modeling flows with a free 4 boundary were successfully solved (Clavin, Lallemand, Pomeau, & Searby, 1988), heat flux distribution (Chopard, & Droz, 1988), domain growth and dynamics (Jacobs, & Masters, 1994), dendritic growth (Plap, & Gouyet, 1997), crowd movement descriptions (Malinetskii, & Stepantsev, 1997). Recently, the theory of cellular automata has been quite promisingly applied to the question of developing self-healing electronic circuits (Koronovsky, Hramov, & Anfinogentov, 1999; Danielli, & Oliveira, 2017; Kasper, 2006).

Cellular automata can solve many circuitry problems using Quantum Dot Cellular Automata (QCA), which implement a new technology with great prospects (Jamal, Tripathi, & Wairya, 2020; Sadhu, Das, De, & Kanjial, 2020; Kassa, & Nagaria. 2016).

Recently, there are many works devoted to finding solutions to the problems of human behavior in various life situations. Especially in natural disasters and other difficult situations (Chen, Y., at all 2020; Sabeur, at all 2015; Verykokou, at all 2018; Doulamis, N., Agrafiotis, P., Athanasiou, G., & Amditis, A. (2017). The results will help save the lives of many people who find themselves in difficult situations, such as fires, floods, etc.

Let's note one more actual application of cellular automata in security systems - data encryption and compression. Also, cellular automata are applicable in the implementation of an effective pattern recognition system. Cellular automata can also be used to solve optimization problems. Often in various fields of activity, there are problems of finding the best option from an unlimited number of possible. As a rule, an exact solution is not required, but a discrete computer is not even able to give an optimal result approximately.

Due to the fact that to describe the values of cells in automata, only Boolean algebra is required, and not real numbers with a large number of decimal places, and the locality of cell rules allows you to create special computers to simulate automata with simpler elements than conventional computers, and with parallel information processing. Firstly, it increases the speed of calculations. Secondly, task programming becomes much easier (you only need to set the rule of the machine in the form of a table of inputs / outputs). Thirdly, the topology of the simulated object in such cellular automaton machines is reproduced by the simulator itself, unlike traditional computers, which undoubtedly has its own advantages in the study of simulated phenomena, making this process more visual.

Now the theory of Cellular Automata is an established scientific discipline with numerous applications in very many fields of science (Berkovich, 1993; Mainzer 2007; Chaudhuri, et al. 1997).

Wolfram mentions more than 10 thousand articles citing his original work on this subject. Being a mathematical object, cellular automata are applicable not only in mathematics. They play an important role as models of spatially distributed dynamical systems, since they initially possess a number of fundamental properties inherent in the physical world: parallelism, uniformity, and local interaction.

Classification of Existing Cellular Automata

Over more than sixty years of the history of cellular automata, a very large number of different variations of cellular automata have been invented and studied, and as a result, a fairly complete systematization of them is a non-trivial task. The existing classification of CA is quite diverse depending on the parameter that defines a particular class.

Of the current classifications of CA, the most famous is the classification of tungsten, based on the analysis of the possible complexity of the behavior of CA (Wolfram, 2002). S. Wolfram classification was the first attempt to classify the rules themselves, rather than the individual types of behavior of the rules.

A positive point in the classification of S. Wolfram is the meaningfulness of the obtained groups, which characterize the possibilities for the dissemination and processing of information. However, this approach also has negative aspects - the complexity of class definition for most CAs.

In addition to the possible complexity of the behavior of the CA, there are other signs by which the CA can be systematized: CA geometry, type of neighborhood, local rules, ways of organizing the automaton in time. Some types of CA are obtained by combining various variant parameters.

According to the spatial characteristics of the CA are divided into one-dimensional and multidimensional depending on the number of spatial coordinates of the automaton lattice. One-dimensional is the class of elementary CA most fully studied and described by Stephen Wolfram.

An elementary CA is a one-dimensional array consisting of finite state machines, each of which can be in two states: 0 or 1, and change its state at a discrete moment of time depending on its state and the state of two nearest neighbors (left and right). All cells update state at the same time.

In practice, the most frequently used multidimensional CA are two-dimensional, less often three-dimensional.

From geometrical characteristics, the neighborhood and isometry are very conventionally allocated. The neighborhood, which consists of cells having a common vertex with a given one, the Moore neighborhood is called. The neighborhood, which consists of cells that have a common side with this, the von Neumann neighborhood is called.

Neighborhood and isometry are interconnected and also depend on time characteristics. Depending on whether the automaton's geometry changes in time, CA are divided into dynamic and static. A wide variety of different options are inherent to the isometric characteristic.

According to the time characteristics of the CA are divided into synchronous, asynchronous, static and dynamic. In synchronous CA, the transition to a new state for all cells is carried out simultaneously by the signal of the global timer. In asynchronous CA, cells transition to a new state in random order.

According to the method of formation of CA recipes laws, this classification distinguishes deterministic (the subsequent state of the cell is uniquely determined by the state of this cell and its closest neighbors at the previous moment in time), probabilistic (the states of the cells at the next moment in time are determined based on some probabilities), generalizing (the rules depend only of the total number of values of neighboring cells). On the basis of homogeneity, CA are divided into homogeneous and inhomogeneous. In homogeneous CAs, the local transition functions and neighborhood indices are the same for each cell. In inhomogeneous CA cells may have different transition functions or different neighborhood indices. Thus, two parameters are involved in the formation of this feature: the principle of the formation of laws-recipes and the method of organizing the neighborhood.

According to the type of state, CA are divided into polygenic, whose elements contain a different set of possible states E_n at different points in time. In practice, cells with equivalent states at each moment of time are used - a linear CA. Linear CA include binary (state states 0, 1), color (from two to 224 states), real (state of cells is described by a set of real variables), CA with memory (memory properties are added to cell xi from its own history), CA without memory (traditional CA). Hyperlogic-based CA are also found in the literature. (Anoprienko, & Konopleva, 2007).

The following sections discuss in more detail some of the above classifications.

ELEMENTARY CELLULAR AUTOMATA

As a relatively simple example, we consider one-dimensional cellular automata. Cells of a one-dimensional automaton are located at integer nodes of the coordinate line. For each cell, neighbors are naturally determined. The neighbors of the nth cell are all cells remote from the given one at a distance of no more than r. The parameter $r \in N$ is called the rank of the cellular automaton. The minimum rank is oneIn this case, there are only two neighbors with numbers $i \pm 1$, and the local neighborhood of the ith cell contains three cells i-1, i and i+1.

The number of cells in an automaton can be infinite or finite. In the latter case, it is assumed that the cells are located at points with coordinates from 0 to (n-1). The homogeneity condition is not fulfilled for the boundary points of this segment, because these points have only one neighbor. Therefore, it is necessary to introduce special conditions, called boundary, which would say how the state of the boundary cells should change. In the most common cases, the state of the boundary cells is either simply fixed (for example, zero boundary conditions), or the so-called cyclic conditions are used, when the segment [0 ... n-1] collapses into a cycle - immediately after the (n-1) -th point there is a zero point.

Another numerical parameter of a one-dimensional cellular automaton is the number of states m. The minimum number is two; such automata are called binary. The states of the binary automaton are denoted as "0" and "1". One-dimensional binary cellular automata of rank 1 are called elementary. Thus, in order to determine an elementary cellular automaton, it is enough to set its system of rules (and boundary conditions, if the number of cells is finite). Since the local neighborhood of any cell of an elementary automaton contains only three cells, each of which can be in two states, then you need to set $2^3 = 8$ rules, for each possible state of the local neighborhood of a point. These rules can be specified either by a formula or by a table.

It is easy to calculate the number of elementary cellular automata. Obviously, this number coincides with the number of logical functions of three variables. This means that there are a total of 256 elementary cellular automata. All these automata can be numbered as follows. The automaton truth table is taken, its last column is written as a string (but from bottom to top), this record is the desired number in the binary number system.

The concept of an elementary automaton was first introduced by the American mathematician Stephen Wolfram in the 1980s, he also conducted a detailed analysis of all 256 automata, as a result of which a classification of cellular automata was proposed based on their type of behavior (Wolfram, 2002):

- Class 1: The result of the evolution of almost all initial conditions is the rapid stabilization of the state and its homogeneity. Any random constructs in such rules quickly disappear;
- Class 2: The result of the evolution of almost all initial conditions is the rapid stabilization of the state, or the occurrence of oscillations. Most random structures quickly disappear under initial conditions, but some remain. Local changes in the initial conditions have a local character on the further course of the evolution of the system;
- Class 3: The result of the evolution of almost all initial conditions is pseudo-random, chaotic sequences. Any stable structures that arise almost immediately are destroyed by the noise surrounding them. Local changes in the initial conditions have a wide, undetectable influence on the course of the entire evolution of the system;
- Class 4: The evolution of almost all rules results in structures that interact in a complex and interesting way with the formation of local, stable structures that can survive for a long time. As a result of the evolution of the rules of this class, some sequences of Class 2 described above can be obtained. Local changes in the initial conditions have a wide, undetectable influence on the course of the entire evolution of the system. Some cellular automata of this class have the Turing universality property.

Of greatest interest from the point of view of this classification are automata of the 3rd and 4th types. The behavior of automata of the third type is chaotic, then if the initial state is completely ordered. An example of an automaton of the 3rd type is an automaton with number 30. The behavior of automata of the fourth type is a mixture of chaotic and deterministic types of behavior - in the process of evolution of an automaton various structures arise that interact in a very non-trivial way. Among elementary automata, the automaton number 110 has such behavior. In particular, it was proved only for this automaton that it is algorithmically universal.

Elementary cellular automata are generalized either by increasing the rank or by increasing the number of states. In both cases, the number of possible machines is extremely large. It is easy to derive a formula for the number of one-dimensional cellular automata of rank r with the number of states m:

$m^{m^{\wedge(2r+1)}}$. For example, there are 2^{32} (approximately 4 billion) binary automata of rank $r = 2$, and 3^{27} (more than 7 trillion) ternary automata of rank $r = 1$. Obviously, the analysis of such automata by a complete search of all variants of the rules is simply impossible.

TWO-DIMENSIONAL CELLULAR AUTOMATA

In a two-dimensional (planar) cellular automaton, the lattice is realized by a two-dimensional array. Adding a second dimension allows you to increase the size of the cell neighborhood.

The cells of two-dimensional automata are located on the plane at points with integer coordinates, i.e., they form a rectangular lattice. The addition of the second dimension allows one to increase the size of the local neighborhood of cells and, accordingly, provides more freedom for constructing cellular automata. A significant difference between two-dimensional automata compared to one-dimensional ones is that in the two-dimensional case several different types of local neighborhoods are possible. Traditionally, two types of local neighborhood are used - von Neumann and Moore.

In the vicinity of von Neumann, the neighbors of a given cell are cells with a common side (common edge). Thus, each cell (with the possible exception of boundary cells) has exactly four neighbors. If the cell coordinates are (i, j), then its neighbors have coordinates (i -1, j) - left neighbor, (i + 1, j) - right neighbor, (i, j-1) upper neighbor (ordinate axis cellular automata are usually considered to be directed downward), (i, j + 1) is the lower neighbor (Figure 1, a - only hatched cells). In the neighborhood of Moore, two cells are adjacent if they have either a common edge or a common vertex. That is, in the local neighborhood of each cell (besides itself) there are eight cells (Figure 1, a - hatched and light gray cells).

Figure 1. Possible options for full coverage of two-dimensional CA from regular polygons

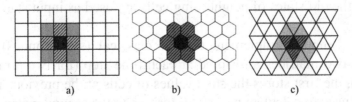

a)   b)   c)

It should be noted that in addition to the self-evident and most often used in practice covering a plane of square components (orthogonal coating), there are only two types of regular polygons that can completely fill two-dimensional space: this is a mosaic of hexagons - a hexagonal coating (Figure 1, b), and a mosaic of equilateral triangles (Figure 1, c). They are also used to construct CA, albeit less frequently compared to the classic orthogonal coating. In addition, despite the narrower spread of the hexagonal coating, it can be considered scientifically proven to be more effective in some specific tasks. (Motornyuk, 2013).

Obviously, the size of the local neighborhood of two-dimensional automata is significantly larger than in the one-dimensional case. This, in particular, means that even for binary two-dimensional automata the number of possible rule sets turns out to be too large (for orthogonal coverage 2^{32} for the von Neumann neighborhood, 2^{512} for the Moore neighborhood) for their complete analysis. Therefore, the main method for studying two-dimensional automata is their synthesis, the selection of a suitable configuration and the development of a set of rules for solving a certain problem (and the subsequent analysis of this particular automaton).

The above definitions of neighborhoods can be parameterized by introducing the rank. We assume that cell c1 enters a local neighborhood of rank r of cell c_0 if c_0 can be reached in c_1 by making at most r transitions. For the von Neumann neighborhood, transitions are performed through the common side of the cells, for the Moore neighborhood either through the common side or through the common vertex (i.e., diagonally).

Note that the most famous example of the implementation of a two-dimensional CA is the game "Life", invented by John Conway in 1970.

SYNCHRONOUS CELLULAR AUTOMATA

Cellular automata are a distributed system of parallel-functioning objects. In synchronous SC (SCA), the state of all cells according to the transition function changes simultaneously by a single external signal (sync pulse). In this case, the old states of neighboring cells are used as input states (Bilan, 2017).

In the computer implementation of classical (synchronous) cellular automata, two matrices are stored at each time step (for two-dimensional automata), the first stores the state values of cells at the previous iteration, which are used to calculate new states written to the second matrix (current states). At the beginning of the next iteration, the first matrix is completely overwritten by the values of the second and the process repeats.

ASYNCHRONOUS CELLULAR AUTOMATA

The presence of a common timer, according to the signals of which the state of all cells is updated (as in the synchronous CA considered above), looks obvious from the point of view of the implementation of the cellular automaton on a computer. However, if we consider cellular automata in the context of modeling, for example, biological systems, then such a synchronous update of the states of all cells is not quite natural.

The process of computer simulation of asynchronous automata (Bilan, 2017), compared with synchronous ones, is simplified as follows: all states are stored in one matrix, automaton cells are sorted in random order, for each of them a new state is calculated, which is immediately written to the state matrix and then used by other cells to update their states. Since the cells of an asynchronous automaton are randomly sorted, the evolution of asynchronous automata is non-deterministic (in particular, the evolution of asynchronous automata is weakly dependent on the initial conditions).

Several different schemes are used to update cells in asynchronous cellular automata:

- for the subsequent update, an arbitrary random cell of the automaton is selected;
- at every step, all cells are randomly selected;
- the search order is drawn up once and is used further at all time steps;
- each cell has its own timer (deterministic or non-deterministic) by the signals of which it is updated.

PROBABILISTIC CELLULAR AUTOMATA

In deterministic CA (DCA), the new state of the cell is uniquely determined by the current state of its neighbors, departing from the same initial conditions, we inevitably get the same evolution. In them, the state of a cell at a subsequent moment in time is uniquely determined by the state of this cell and its nearest neighbors at a previous moment in time. The values of all parameters of the model are determined by determinate values (i.e., each parameter corresponds to a specific integer, real or complex number or the corresponding function). This method corresponds to the full certainty of the parameters.

Diametrically opposite in terms of certainty are CA operating on the basis of a probabilistic rule, when the same current situation can lead to several different results with a given probability of each of them. Stochastic - when the values of all or individual model parameters are determined by random variables given by probability densities. In the literature, the cases of normal (Gaussian), uniform and exponential distribution of random variables are most fully investigated.

Cellular automata, in which the states of cells at the next instant of time are oriented based on certain probabilities, are called probabilistic (or stochastic) cellular automata. In classical probabilistic cellular automata, transition rules are abstract in nature and are not uniquely connected with real processes occurring in a simulated system. In such automata, when modeling a process for any cell, a random quantity Q is generated by a random quantity sensor (0 < Q < 1), which is compared with the probability W of the implementation of this process. If Q < W, then the process is being implemented.

Such CA include the reaction lattice gas method, the Monte Carlo direct stimulation method, and the probabilistic CA method using the Monte Carlo exercise.

In probabilistic CA, it is necessary to establish a set of probabilities of a change in the state of the cell, which show what is the probability of the transition of the ith component from the state at the nth time moment to the state at the next n + 1st time moment, provided that the states of its nearest neighbors in n-th moment of time took certain values.

The following methods of composition of substitutions are usually used in the local stochastic CA operator: superposition, random selection of one of the substitutions, the use of substitutions in random order (Kalgin, 2011).

Probabilistic CA, due to their uncertainty, are most often used in low-level discrete modeling of natural (primarily biological) systems (Ershov, & Kravchuk, 2014; Sinyachkin, Petrov, & Ershov, 2015).

CONCLUSION

This chapter discusses the history of the creation and formation of the theory of CAs. The emphasis is made on existing CAs that describe the behavior of people in various situations. Modern models made it possible to solve the problems of moving a group of people, taking into account the analysis of the features of their behavior during movement. The existing results made it possible to choose a scientific direction that is aimed at describing the behavior of living organisms and their interaction based on ACA. This direction is developing on the basis of studies of the behavior of active cells in the environment of ACA and expands their modern classification.

REFERENCES

Aladyev, V. (1974). Survey of research in the theory of homogeneous structures and their appllctions. *Mathematical Biosciences*, *15*, 121–154. doi:10.1016/0025-5564(74)90088-1

Anoprienko, A., & Konopleva, A. (2007). The experience of using hypercodes in modeling cellular automata. Scientific works of Donetsk National Technical University. A series of "Problems of modeling and design automation of dynamic systems" (MAP-2007). Donetsk. *DonNTU*, *6*(127), 220–227.

Berkovich, S. (1993). *Cellular automata as a model of reality*. Per.

Berthold, O. (2009). *Computational universes*. Berlin: Humboldt Universitat zu Berlin, Institut fur Informatik.

Bilan, S. (2017). *Formation Methods, Models, and Hardware Implementation of Pseudorandom Number Generators: Emerging Research and Opportunities*. IGI Global.

Burks, A. (Ed.). (1970). *Essays on Celluar Automata*. University of Illinois Press.

Chaudhuri, P. P., Chowdhury, D. R., Nandi, S., & Chattopad-hyay, S. (1997). *Additive cellular automata, theory and applications* (Vol. 1). John Wiley & Sons.

Chen, Y., Wang, C., Li, H., Yap, J. B. H., Tang, R., & Xu, B. (2020). Cellular automaton model for social forces interaction in building evacuation for sustainable society. *Sustainable Cities and Society*, *53*, 101913. doi:10.1016/j.scs.2019.101913

Chopard, B., & Droz, M. (1988). Cellular automata model for heat conduction in a fluid. *Physics Letters. [Part A]*, *126*(8/9), 476–480. doi:10.1016/0375-9601(88)90042-4

Clavin, P., Lallemand, P., Pomeau, Y., & Searby, G. (1988). Simulatoin of free boundaries in flow system by lattice-gas m odels. *Journal of Fluid Mechanics*, *188*, 437–464. doi:10.1017/S0022112088000795

Csahyk, Z., & Vicsek, T. (1995). Lattice gas model for collective biological motion. *Physical Review*, *52*(5), 5297–5303. PMID:9964028

Danielli, L., & Oliveira, G. (2017). A cellular automata ant memory model of foraging in a swarm of robots. *Applied Mathematical Modelling*, *47*, 551–572. doi:10.1016/j.apm.2017.03.021

Doulamis, N., Agrafiotis, P., Athanasiou, G., & Amditis, A. (2017). Human object detection using very low resolution thermal cameras for urban search and rescue. *Proceedings of the 10th International Conference on PErvasive Technologies Related to Assistive Environments*, 311-318. 10.1145/3056540.3076201

Ershov, N., & Kravchuk, A. (2014). Discrete modeling using stochastic cellular automata. *Bulletin of the Peoples' Friendship University of Russia. Series: Mathematics, Computer Science, Physics, 2*, 361-364.

Flake, G. W. (1998). *The computational beauty of Nature*. MIT Press.

Frisch, U., D'humières, D., Hasslacher, B., Lallemand, P., Pomeau, Y., & Rivent, J. P. (1987). Lattice gas hydrodynamics in two and three dimensions. *Complex Systems, 1*, 649–707.

Gernert, D. (1989). Cellular automata and the concept of space. *Proceedings of the Workshop on Parallel Processing: Logic, Organization, and Technology (WOPPLOT 89)*, 565, 94–102.

Jacobs, D. J., & Masters, A. J. (1994). Domain growth in one-dimens ional diffusive lattice gas with short-range attraction. *Physical Review A., 49*(4), 2700–2710. PMID:9961535

Jamal, K. S., Tripathi, D., & Wairya, S. (2020). *Design of Full Adder with Self-checking Capability Using Quantum Dot Cellular Automata. In Advances in VLSI, Communication, and Signal Processing*. Springer.

Kalgin, K. V. (2011). Domain specific language and translator for cellular automata models of physicochemical processes. Berlin: Springer. *Lecture Notes in Computer Science, 6873*, 172–180. doi:10.1007/978-3-642-23178-0_14

Kasper, S. (2006). Using cellular automata and gradients to control self-reconfiguration. *Robotics and Autonomous Systems, 54*(2), 135–141. doi:10.1016/j.robot.2005.09.017

Kassa, S. R., & Nagaria, R. K. (2016). An Innovative Low Power Full Adder Design in Nano Technology Based Quantum Dot Cellular Automata. *Journal of Low Power Electronics., 12*, 1–5. doi:10.1166/jolpe.2016.1431

Kauffman, S. A. (1995). *At home in the Universe: The search for laws of self-organization and complexity.* Oxford University Press.

Koronovsky, A. A., Hramov, A. E., & Anfinogentov, V. G. (1999). Phenomenological model of electron flow with a virtual cathode. *Proceedings of the RAS. Ser. Physical, 63*(12), 2355-2362.

Llachinski, A. (2002). *Cellular automata: A discrete Universe* (2nd ed.). World Scientific Publ. Co.

Mainzer, K. (2007). *Thinking in complexity. The computational dynamics of matter, mind, and mankind.* Springer.

Malinetskii, G. G., & Stepantsev, M. E. (1997). Modeling crowd movement with the help of cellular automata. News of universities. Ser. *Izvestiya Vyssih Ucebnyh Zavedenij. Prikladnaya Nelinejnaya Dinamika, 5*, 75–79.

Motornyuk, R. L. (2013). *Computer-aided methods for identifying images of moving objects based on cellular automata with a hexagonal coating.* Dissertation for the degree of candidate of technical sciences (UDC 004.932: 519.713 (043.3)). Kiev: SUIT, 2013

Neumann, J. (1966). *Theory of Self-Reproducing Automata: Edited and completed by A. Burks.* University of Illinois Press.

Sabeur, Z., Doulamis, N., Middleton, L., Arbab-Zavar, B., Correndo, G., & Amditis, A. (2015) Multi-modal computer vision for the detection of multi-scale crowd physical motions and behavior in confined spaces. *International Symposium on Visual Computing*, 162-173. 10.1007/978-3-319-27857-5_15

Sadhu, A., Das, K., De, D., & Kanjial, M. R. (2020). Area-Delay-Energy aware SRAM memory cell and M× N parallel read/write memory array design for quantum dot cellular automata. *Microprocessors and Microsystems, 72*(102944), 1–21. doi:10.1016/j.micpro.2019.102944

Schiff, J. L. (2008). *Cellular automata. A Discrete View of the World.* John Wiley & Sons Inc.

Sinyachkin, A., Petrov, M., & Ershov, N. (2015). Instrumental environment for low-level modeling based on stochastic block cellular automata. In *Information and telecommunication technologies and mathematical modeling of high-tech systems, Materials of the All-Russian Conference with international participation.* PFUR.

Smith, A. III. (1976). Introduction to and survey of polyautomata theory. In *Automata, Languages, Development.* North Holland Publishing Co.

Toffoli, T. & Margolus, N. (1991). *Machines of cellular automata.* Per.

Verykokou, S., Ioannidis, C., Athanasiou, G., Doulamis, N., & Amditis, A. (2018). 3D reconstruction of disaster scenes for urban search and rescue. *Multimedia Tools and Applications, 77*(8), 9691–9717. doi:10.100711042-017-5450-y

Vollmar, T. (1977). Cellular spaces and parallel algorithms, and introductory survey. In M. Feilmeier (Ed.), *Parallel ComputationParallel Mathematics* (pp. 49–58). North Holland Publishing Co.

Wiener, N., & Rosenbluth, A. (1946). The mathematical formulation of the problem of conduction of impulses in a network of connected excitable elements, specifically in cardiac muscle. *Archivos del Instituto de Cardiologia de Mexico, 16,* 205–265. PMID:20245817

Wolfram, S. (1983). Statistical mechanics of cellular automata. *Reviews of Modern Physics, 55*(3), 601–644. doi:10.1103/RevModPhys.55.601

Wolfram, S. (1986). *Appendix of Theory and Applications of Cellular Automata.* World Scientific.

Wolfram, S. (1986a). Cellular automation Fluids. *Journal of Statistical Physics, 45*(3-4), 471–526. doi:10.1007/BF01021083

Wolfram, S. (2002). *A new kind of science.* Wolfram Media.

Zuse, K. (1969). *Rechnender Raum.* Vieweg. doi:10.1007/978-3-663-02723-2

Chapter 2
Modeling of Dynamic Processes

ABSTRACT

The chapter describes the existing approach to modeling dynamic systems. The rules of the transfer of properties and conditions from cell to cell in cellular automata of various organizations are considered. The basic cell structures are presented in the transfer of only states, as well as properties of cell activity and states. The options are considered when the cell itself selects a cell among the cells in the neighborhood that will become active in the next time step. Also is considered is the option when the cell analyzes the state of neighboring cells and, based on the results of the local state function, makes a decision about the transition to the active state or not. An embodiment of a cell for transmitting an active state is described, only to cells with a given local logical function. Cell structures and their CAD models are constructed.

INTRODUCTION

Time is an integral characteristic of our life and all material objects on earth. Over time, all properties and shapes of objects change. Such a change is called a dynamic process. If nothing happens with the object and properties over time (their quantitative characteristics do not change), then the process is called static. However, in the modern world, static processes can be present only at short time intervals, which can be achieved by eliminating all external influences on the object. Depending on the stability of the object and the

DOI: 10.4018/978-1-7998-2649-1.ch002

strength of external influences, all objects can change at different times, which indicates the presence of dynamic processes. In modern studies, dynamic processes are described by differential equations or systems of differential equations. However, they are not always suitable for describing the dynamics of the behavior of many objects. One of the solutions for describing the dynamics of the behavior of objects is the use of cellular automata.

The chapter describes the existing approach to modeling dynamic systems. The rules of the transfer of properties and conditions from cell to cell in cellular automata of various organizations are considered. The basic cell structures are presented in the transfer of only states, as well as properties of cell activity and states. The options are considered when the cell itself selects a cell among the cells in the neighborhood that will become active in the next time step. Also is considered is the option when the cell analyzes the state of neighboring cells and, based on the results of the local state function, makes a decision about the transition to the active state or not. An embodiment of a cell for transmitting an active state is described, only to cells with a given local logical function. Cell structures and their CAD models are constructed.

EXISTING APPROACHES TO MODELING DYNAMIC PROCESSES

Everything in the modern world is changing in space and time. The process of changing any properties of objects or processes in space and time is called a dynamic process. Various dynamic processes constantly occur in the world. The most understandable dynamic process for humans is the movement of an object. A moving object is precisely determined by a change in its location in space at various points in time.

However, the coordinates of objects do not change in space in many dynamic processes. In this case, other quantitative characteristics of the properties of the object may change or the properties of the object itself may change. New properties may appear or existing properties may disappear.

The dynamic process is characterized by two factors:

- the structure of the system of elements that forms the object and the initial states of the elements of the system;
- the interaction of several objects.

The structure of the system determines the order of interaction of its elements that implement a specific function. In such a system, the paradigm of the dynamic process is embedded.

In the literature, a dynamic system is defined as a set of elements for which a functional relationship between time and position in the phase space of each element of the system is given (Evlanov, 1972; Giunti, & Mazzola, 2012; Galor, 2010).

The phase space of the system describes all possible states in which the system may be. A sequence of states describes the behavior of the system in time and space, and a description of the sequence of states of the system in time from the initial one represents the evolution of the system in time.

Each dynamic process is determined by an initial and final time count. The amount of time between the initial and final time samples determines the life cycle of a dynamic process. During the life cycle (LC), various changes in the quantitative characteristics of various properties of objects occur. Thus, LC can be defined by a variety of states of systems and elements that are affected by a dynamic process (DP).

The length of the life cycle of all DPs is determined by the energy charge or the amount of energy required to perform the DP.

The amount of energy and energy charge of a dynamic process is not always determined by physical energy such as, for example, the amount of fuel to move a car 100 km. There are many dynamic processes whose LCs are determined by the structure of the system and its initial settings. For example, a huge number of people live on earth, but each of them lived a certain amount of time.

It is known that the system is described by many elements and the relationships between them. If the system is not affected by external influences, then the length of its life cycle is determined by its structure, otherwise the length of the life cycle of the system changes upwards (positive external influences) or decreases (negative external influences).

Thus, the LC system can be described by the following model

$$LC_S = N,\ SN,\ BS,\ OF,$$

where

N – number of system elements;
SN – system of connections between elements;
BS – initial state of the system;

OF – many external influences on the system.

If you remove the component OF = 0, then the system will go through its life cycle, which is determined by its internal structure. The life cycle of such a system will be determined by many states in time $LC=\{S\}$. However, with the same structure of the system, but with different initial states, the system has different LCs.

The process of functioning of a system in time is called evolution.

In nature, there are systems that die at the end of their life cycle. Moreover, during the life cycle, they generate new systems that become more advanced and have a large number of functional capabilities. Such systems can themselves generate new, more advanced systems.

Systems that generate their own kind can themselves reproduce new systems, but can only reproduce in tandem with another system. Each new born system as a rule should have better stable properties than the systems that generated it. When a new system is formed by one system, the new system transfers the properties from the old system, as well as the experience gained during the evolution of the old system (during its life cycle).

In the case when several old systems reproduce the new system, the new system was transferred the properties and structure from the old systems, as well as the experience gained by the old systems. In addition, the new system can acquire properties that are the result (option) of the meeting of the properties of two old systems.

It is also possible that the properties of one of the old systems are transferred to the new system.

Thus, the evolution of systems of similar structure continues through the transfer of properties and accumulated experience.

To date, many methods have been developed that are aimed at studying and modeling the processes of self-reproducing systems (Byl, 1989; Langton, 1984; Neumann, 1966; Wolfram, 1986; Tempesti, Mange, & Stauffer 2009; Prokopenko, 2013; Yates, 2012; Burguillo, 2017; Murata, & Kurokawa, 2012). However, little attention is paid to modeling the processes of reproducing new systems by several systems while maintaining the structure and properties of old systems. CA are well suited for modeling such processes.

TRANSFER OF PROPERTIES AND STATES IN CELLULAR AUTOMATA

CA are arranged in such a way that at each time step the cells change their state. Cell states vary depending on the selected neighborhood structure and local state function. A cell may not change its state due to the many states of cells in its neighborhood. In such CA, the cell has the property to realize the selected local state function (LSF). LSF arguments are signals of the state of cells in the neighborhood.

In fact, in such known cellular automata, the transfer of properties and conditions from cell to cell or from multiple cells to one cell does not occur. There is a possibility for a specific set of LSF, in which the state of logical "1" is transferred to neighboring cells (operation "OR") from one cell that has a logical state of "1", as well as from neighborhood cells to its own cell (operation "AND"). In such cases, the number of cells with an ones state increases (function "OR") or decreases (operation "AND"). Examples of the functioning of such CA on Figure 1 are presented.

An analysis of the figures shows that such CA have a short operating time.

Figure 1. Examples of the functioning of CA with the transfer of states from cells to cells

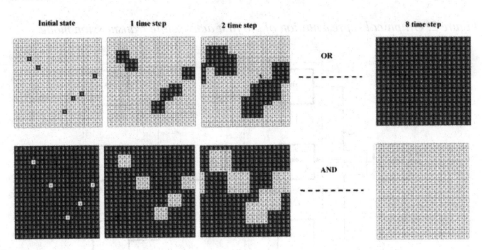

Another approach is to transfer properties from one cell to only one cell in the neighborhood. In such CA, the cell that transfers the property to functioning is active. Only an active cell can perform LSF.

The principles of the transfer of active properties are as follows.

A cell of such an automaton has two states: information state and active state.

Only a cell that is in an active state can perform LSF. A cell can be in an active state in only one time step. At the end of the next time step, the active cell itself can select one of the cells in the neighborhood, which in the next time step will become active, or maybe vice versa, one of the cells in the neighborhood of the active cell at the next time step goes into the active state.

Thus, the transfer of the active state by one active cell can be carried out in two modes (Bilan, 2017; Bilan, Bilan, & Bilan, 2015; Bilan, Bilan, Motornyuk, Bilan, & Bilan, 2016).

1. The mode of transmission of an active state from an active cell to a passive cell in a neighborhood.
2. The mode of transition of a passive cell belonging to the active cell in the neighborhood of the active.

Figure 2. Graphical representation of the first active state transmission mode

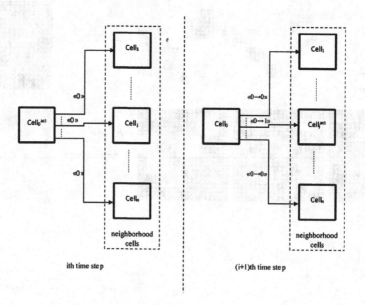

ith time step

(i+1)th time step

The first mode of active state transmission is graphically described in Figure 2.

In the first mode, the active cell performs a local transmission function (LTF) of the active state. Arguments of LTF are signals of informational state of the neighborhood cells. According to the LTF and the state of the neighborhood cells, the active cell selects one of the neighborhood cells, which at the next instant will go into the active state. At the corresponding active output of the active cell ($Cell_0^{act}$), a logical "1" signal is generated, which is transmitted to the active input of the selected neighborhood cell.

The second mode of transmission of the active state on Figure 3 is graphically presented.

In the second mode, all cells in the neighborhood of the active cell are in a state of readiness for transition to the active state, since at their active inputs there is a single signal from the active output of the active cell. Each cell in the neighborhood of the active cell analyzes the signals from the cells of its neighborhood, or from other cells and makes a decision on its transition to the active state. For example, a neighborhood cell is equipped with four active inputs that are connected to the corresponding active outputs of neighborhood cells. A cell goes into an active state if, for example, a logical "1" signal is present at its third active input (Figure 3).

Very important for both modes is the choice of the LTF, which should not be unidirectional. In this mode, the active cell performs LTF, the result of which is the appearance of a logical "1" on one of the active outputs, which indicates the active cell in the next time step.

As a rule, all active cells analyze cells belonging to its neighborhood. In this situation, the neighborhood cells must have different states for all CA cells. A function, wich indicates the number of the active output, have good showing, i.e. LTF analyzes the first two cells of the von Neumann neighborhood or the first three cells of the Moore neighborhood (Bilan, 2017; Bilan, Bilan, & Bilan, 2015; Bilan, Bilan, Motornyuk, Bilan, & Bilan, 2016; Bilan, Bilan, & Bilan, 2017). An analysis of the use of majority functions was also conducted. However, structures with two clock cycles of active signal transmission showed good results. (Bilan, 2017; Bilan, Bilan, & Bilan, 2015; Bilan, Bilan, Motornyuk, Bilan, & Bilan, 2016; Bilan, Bilan, & Bilan, 2017). A complete set of the LTFs that show high efficacy has not yet been determined. The effectiveness of each LTF is determined by the target membership of the modeling task.

These modes of transmission of the active state sometimes lead to the fact that the active signal can propagate in only one direction. This is due to an

Figure 3. Graphical representation of the second active state transmission mode

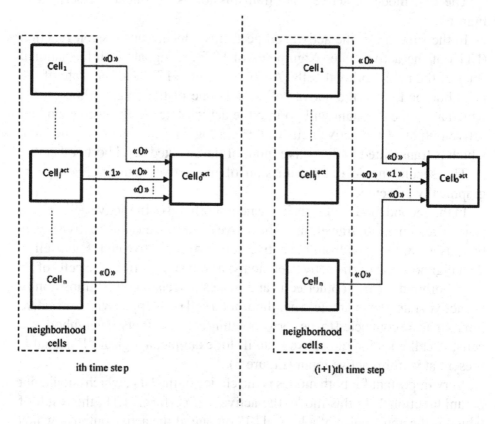

ineffective choice of the LTF. Therefore, another mode is used, which is to transfer the active state according to the results of the LSF.

An active signal can only propagate through cells that perform a specific LSF (Belan, & Belan, 2012; Belan, & Belan, 2013). For example, the cells that are in the logical "1" state can be active cells (Figure 4).

Figure 4. The evolution of CA during the transfer of an active state for cells with two neighboring cells having a logical "1" state

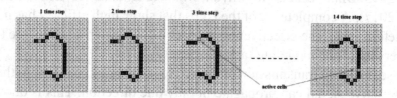

When transferring properties and states, the initial active cell on the CA must always be specified. For the case presented in Figure 4, the initial active cell coincided with a cell that implements the necessary LSF. However, the initial cell may not always coincide with the cell that implements the selected LSF. In this case, the active signal propagates through all the cells of the CA until the active signal reaches a cell that has a single state when a given LSF is performed (Figure 5).

Figure 5. An example of the evolution of the CA shown in Figure 1 with a different arrangement of the initial active cell

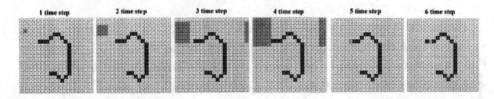

Initially, active cells propagate throughout all CA cells. When a CA cell becomes active with a given LSF, the active signal begins to propagate only through cells with a given LSF. This mode has shown high efficiency in the description and analysis of images (Belan, & Belan, 2012; Belan, & Belan, 2013).

The considered modes realize only the transfer of the active state from the active cell to the neighboring cell, which will become active in the next time step. In this case, LSF and internal cell structures can also be transmitted from space to cell in the cell in CA.

There are two ways to transfer LSF:

- reprogram the cell;
- rebuild the internal structure of the cell.

The first mode requires the use of a programmable device that performs all LSF. The second mode requires a specialized unit, which consists of many circuits that implement all possible LSF or the cell must contain FPGAs for reprogramming LSF schemes. An example of the evolution of the functioning of CA with the transfer of LSF by active cells in Figure 6 is presented.

Figure 6. An example of the evolution of CA during the transmission of LSF (XOR, AND) by two active cells. All CA cells perform OR local state function for both CA.

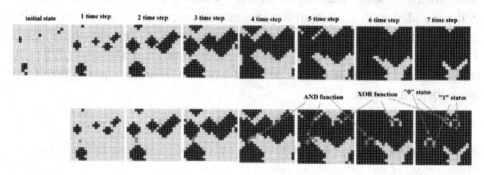

Evolution (Figure 6) shows that the CA operates for a long time, which is longer than the transition time of all cells to the logical "1" state when OR local state function is performed by all CA cells. At the seventh time step, there are more zero states on the second inhomogeneous CA than on a homogeneous CA with one OR local state function. This is determined by the AND function, which in the presence of a zero state constantly holds the zero state of the cell. However, the remaining cells after transition to a single state do not change their state. If cells with AND and XOR functions transfer their properties further to neighboring cells with each subsequent time step, the state of the CA will change significantly, which increases the LC of the CA.

Variants are also possible when two active cells with different LSF are found in the same cell. In this case, a new cell is formed that performs a new LSF (Bilan, Bilan, & Bilan, 2017). This LSF is determined by the following model

$$LFS_{new} = f(LFS_1, LFS_2).$$

The function f() is defined in different ways. Moreover, the limited function is determined by a limited set of LSF for each active cell. The principles of functioning of such CA will be discussed in more detail below.

The number of transferred properties and conditions by active cells is limited by the cell structure. However, the most unforeseen states of the CA as a whole are possible, which arise as a result of the functioning of active cells at a certain number of time steps. Such states of the CA require additional studies.

Hardware Implementation of Cellular Automata With the Transfer of Properties and States

The software implementation of the CA limits the speed of performing computational operations by each cell, and also limits the rate of transfer of properties and states from cell to cell. The hardware implementation of the CA allows improving the performance indicator. Moreover, the uniformity of the structure increases its reliability.

The simplest CA structure consists of homogeneous cells and bonds between cells that organize a neighborhood for each CA cell. Each cell functions according to a predetermined algorithm, according to which the cell takes a certain state at each time step (Figure 7).

Figure 7. Simple two-dimensional SCA cell algorithm

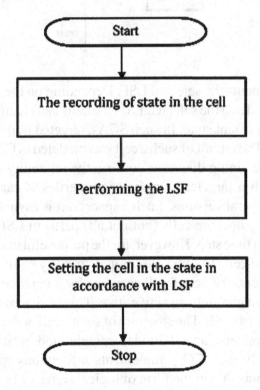

A cell in such an SCA at each time step goes into the state determined by the LSF. The state of the cell also depends on the LSF arguments, which are state signals from the outputs of the cells in the neighborhood. In fact, in such an SCA, the properties of the cells are not transmitted. Only states are transmitted depending on the conditions of all SCA cells. The cell structure of such a CA is described in many works (Bilan, 2017; Bilan, Bilan, & Bilan, 2017). It consists (Figure 8) of a memory element (flip-flop) and a combinational circuit (CC), which controls the state of the flip-flop T.

Figure 8. Functional diagram of a two-dimensional simple SCA cell

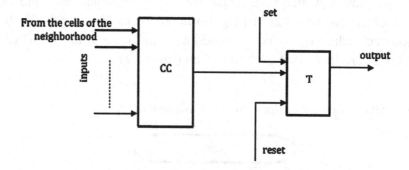

The CC implements the selected LSF. Depending on the state of the cells in the neighborhood and the selected LSF, the cells can change their state, but can remain in the current state. In such SCAs, targeted transfer of properties is not performed. The circuit of such a cell was modeled in CAD Active-HDL (Figure 9), and the timing diagrams confirm the reliability of operation.

A CA in which a targeted transfer of properties is carried out can be implemented in several versions. Such a spacecraft is asynchronous. In it, at each time step, one or more cells (but not all) perform LSF. Such cells are active at the current time step. However, for the purposeful transfer (obtaining) of properties in the cell structure, additional circuitry solutions are used. In the first embodiment, the active cell searches for a neighboring cell, which at the next time step goes into an active state. This next active cell in the next time step will perform LSF. The structure of such a cell is shown in Figure 10.

Such a cell contains an additional combination scheme (CC_{act}) that implements LSF. In fact, CC_{act} implements n functions whose values are formed at the outputs (Y_i) in the form of logical signals «1» or «0». The cell also contains an additional flip-flop (T_{act}), which stores the state of activity. T_{act} has a logical "1" state, then the cell is active, otherwise the cell is inactive. Logical signal "1" is formed at those outputs Y that are connected to the active inputs of neighboring cells, which become active in the next time step.

Figure 9. CAD simulation of an SCA cell that implements a XOR function for the von Neumann and the Moore neighborhoods

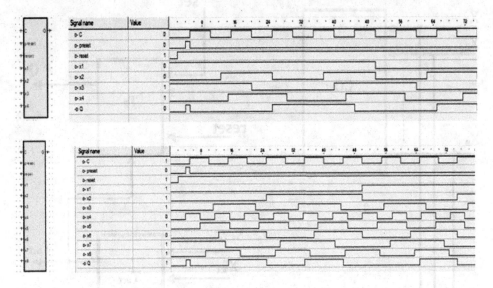

LSF arguments are the values of information signals of states from the information outputs Q of neighboring cells. Neighboring cells organize a neighborhood of the active cell. Also, the first flip-flop T_1 performs LSF only when the second flip-flop T_{act} has a logical state of "1". The second flip-flop T_{act} goes into a logical "1" state, when a logical "1" signal from the output of a neighboring active cell comes to its active input X_{act}.

The circuit was modeled in CAD Active-HDL for the XOR local state function and the neighborhood of Moore (Figure 11). The following functions were selected as LSF, which are represented by the following system of logical functions.

Figure 10. A cell structure that selects an adjacent cell to transmit an active state

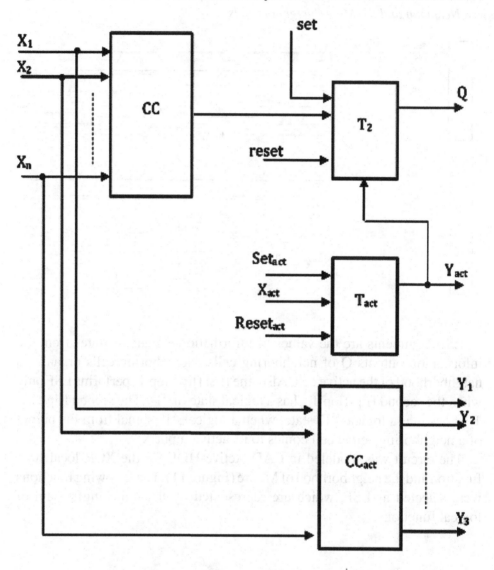

$$\begin{cases} Y_1 = \overline{X_1} \overline{X_2} \overline{X_3} \\ Y_2 = \overline{X_1} \overline{X_2} X_3 \\ Y_3 = \overline{X_1} X_2 \overline{X_3} \\ Y_4 = \overline{X_1} X_2 X_3 \\ Y_5 = X_1 \overline{X_2} \overline{X_3} \\ Y_6 = X_1 \overline{X_2} X_3 \\ Y_7 = X_1 X_2 \overline{X_3} \\ Y_8 = X_1 X_2 X_3 \end{cases} \qquad (1)$$

To select the next active cell, the signals of only three cells of the neighborhood are used since they can form eight different combinations of logical "0" and "1". Also, if the cell is active, then the cell performs LSF, as shown in timelines. The logic signal "1" is formed on one of the active outputs.

Figure 11. CAD modeling of the ACA cell with the transfer of the active state to the neighboring cell according to the LTF represented by the system (1)

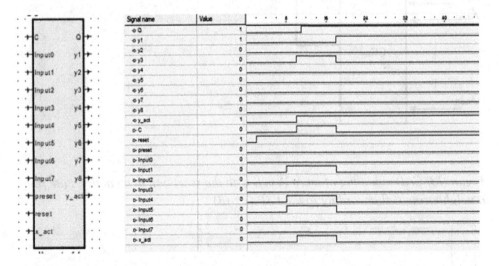

A CA in which an active state is received from one of the cells in a neighborhood contains cells that have only one active output. The cell structure of such a CA on Figure 12 is shown.

Figure 12. Functional diagram of a cell that receives an active state from cells in a neighborhood

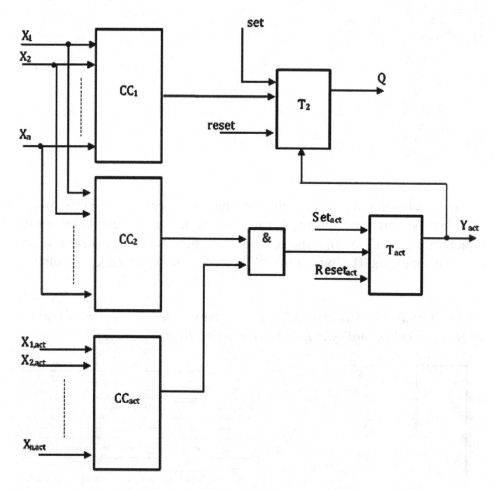

Figure 13. CAD modeling of an ACA cell with the reception of an active state from a neighboring cell according to the LSF (2), (3)

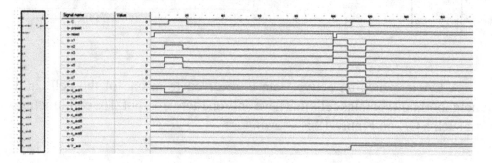

The second combinational circuit (CC_2) and the active combinational circuit (CC_{act}) analyze the information and active signals from the cells of the neighborhood and decide on the transition of the cell to the active state at the next time step. Instead of CC_2, CC_1 can be selected. Alternatively, instead of CC_{act}, OR gate can be used.

This scheme is modeled in CAD Active-HDL for XOR local state function and the neighborhood of Moore. How CC_2 function selected

$$LTF_{CC_2} = \overline{X_1 X_2 X_3 X_4 X_1 X_2 X_3 X_4}, \qquad (2)$$

Figure 14. Functional diagram of a cell that transmits an active signal by analyzing the state of cells in a neighborhood

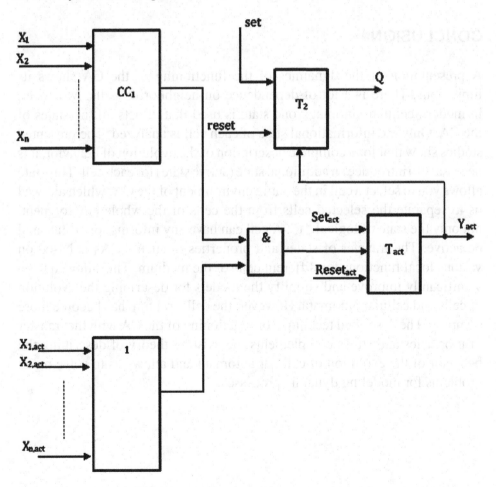

and how CC_{act}

$$LTF_{act} = \overline{X_1 X_2 X_3 X_4 X_5 X_6 X_7 X_8}.\tag{3}$$

The circuit of such a cell was modeled in CAD Active-HDL (Figure 13), and the obtained time diagrams showed high reliability.

In the mode of operation of the cell to transfer the active state according to the results of the LSF, the cell has the same structure as the cell in the previous mode. In this case, K_{act} is an OR gate (Figure 14).

In such a cell, the first T_1 trigger is used only to store the image information state. Only the active state is transmitted from cell to cell with the same eigenstates and equal results of LSF. This mode is used to read information or count cells with specified properties and conditions of cells in the neighborhood.

CONCLUSION

A presentation of the dynamics of the functioning of the CA shows its limitations. There is a lot of dependence on neighboring cells. Moreover, in modern cellular automata, one state is used that affects all the states of the CA. Only the informational state of each cell is analyzed. The presented studies show that for a complete description of the evolution of behavior, it is necessary to introduce an additional state (active state) for each cell. This state allows you to select a cell in the entire environment of the CA, which allowed us to separate the selected cells from the cells of the whole environment, not only the state of logical "1". A cell can have any informational state and be active. The transfer of states and properties in such ACAs is based on various local functions for different cells of the medium. This allowed us to significantly improve and simplify the models for describing the evolution of cells and cellular automata. However, the cell structure has become more complex. The described technique of functioning of the CA with the transfer of properties and states complements the existing theory of describing the behavior of the evolution of cellular automata and allows us to solve many problems for modeling dynamic processes.

REFERENCES

Belan, S., & Belan, N. (2012). Use of Cellular Automata to Create an Artificial System of Image Classification and Recognition. *LNCS*, *7495*, 483–493.

Belan, S., & Belan, N. (2013). Temporal-Impulse Description of Complex Image Based on Cellular Automata. In PaCT2013 (LNCS, Vol. 7979, pp. 291-295). Springer-Verlag.

Bilan, S. (2017). *Formation Methods, Models, and Hardware Implementation of Pseudorandom Number Generators: Emerging Research and Opportunities*. IGI Global.

Bilan, S., Bilan, M., & Bilan, S. (2015). Novel pseudorandom sequence of numbers generator based cellular automata. *Information Technology and Security*, *3*(1), 38–50.

Bilan, S., Bilan, M., & Bilan, S. (2017). Research of the method of pseudo-random number generation based on asynchronous cellular automata with several active cells. *MATEC Web of Conferences, 125*, 1-6. 10.1051/matecconf/201712502018

Bilan, S., Bilan, M., Motornyuk, R., Bilan, A., & Bilan, S. (2016). Research and Analysis of the Pseudorandom Number Generators Implemented on Cellular Automata. *WSEAS Transactions on Systems*, *15*, 275–281.

Burguillo, J. C. (2017). Self-organizing Coalitions for Managing Complexity: Agent-based Simulation of Evolutionary Game Theory Models using Dynamic Social Networks for Interdisciplinary ... Complexity and Computation. Springer.

Byl, J. (1989). Self-reproduction in small cellular automata. *Ibid*, *34*, 295–300.

Evlanov, L. G. (1972). *Control of dynamic systems*. Nauka.

Galor, O. (2010). *Discrete Dynamical Systems*. Springer.

Giunti, M., & Mazzola, C. (2012). Dynamical systems on monoids: toward a general theory of deterministic systems and motion. In Methods, Models, Simulations and Approaches Towards a General Theory of Change. World Scientific. doi:10.1142/9789814383332_0012

Langton, C. G. (1984). Self-replication in cellular automata. *Physica D. Nonlinear Phenomena*, *10*(1-2), 135–144. doi:10.1016/0167-2789(84)90256-2

Murata, S., & Kurokawa, H. (2012). *Self-Organizing Robots (Springer Tracts in Advanced Robotics)*. Springer. doi:10.1007/978-4-431-54055-7

Neumann, J. (1966). *Theory of Self-Reproducing Automata: Edited and completed by A. Burks*. University of Illinois Press.

Prokopenko, M. (Ed.). (2013). *Advances in Applied Self-Organizing Systems (Advanced Information and Knowledge Processing)* (2nd ed.). Springer. doi:10.1007/978-1-4471-5113-5

Tempesti, G., Mange, D., & Stauffer, A. (2009). *Self-replicating and cellular automata*. In R. A. Meyers (Ed.), *Encyclopedia of Complexity and Systems Science* (pp. 8066–8084). Springer-Verlag. doi:10.1007/978-0-387-30440-3_477

Wolfram, S. (1986). *Appendix of Theory and Applications of Cellular Automata*. World Scientific.

Yates, F. E. (Ed.). (2012). *Self-Organizing Systems: The Emergence of Order (Life Science Monographs)*. Springer.

Chapter 3
Models and Paradigms
of Cellular Automata
With One Active Cell

ABSTRACT

The chapter describes the basic models and paradigms for constructing asynchronous cellular automata with one active cell. The rules for performing local state functions and local transition functions are considered. The basic cell structures during the transmission of active signals for various local transmission functions are presented. The option is considered when the cell itself selects among the cells in the neighborhood of the cell, a cell that will become active in the next time step, and also the structure with active cells under control is considered. The analysis of cycles that occur in cellular automata with one active cell is carried out, and approaches to eliminating cycles are formulated. Cell structures are constructed and recommendations for their modeling in modern CAD are formulated.

INTRODUCTION

The chapter describes the basic models and paradigms for constructing asynchronous cellular automata with one active cell.. The rules for performing local state functions and local transition functions are considered. The basic cell structures during the transmission of active signals for various local transmission functions are presented. The option is considered when the

DOI: 10.4018/978-1-7998-2649-1.ch003

cell itself selects among the cells in the neighborhood of the cell a cell that will become active in the next time step, and also the structure with active cells under control is considered. The analysis of cycles that occur in cellular automata with one active cell is carried out, and approaches to eliminating cycles are formulated. Cell structures are constructed and recommendations for their modeling in modern CAD are formulated.

MODEL OF A CELLULAR AUTOMATA WITH A SINGLE ACTIVE CELL

Cellular automata with one active cell belong to the ACA family, in which only one cell at a time changes its state. In this case, at each subsequent moment of time, the cell, which is in a predetermined neighborhood with the active cell at the current time, changes its state.

An active cell is a cell that at the current time can perform a local state function (LSF). In accordance with LSF, an active cell forms its new information state or remains in its current state. Also, an active cell (AC) implements a local transition function (LTF) to select AC at the next time (Bilan, 2017; Bilan, Bilan, & Bilan, 2015; Bilan, Bilan, & Bilan, 2017; Bilan, et al 2016).

Figure 1. Block diagram of an ACA cell with one active cell

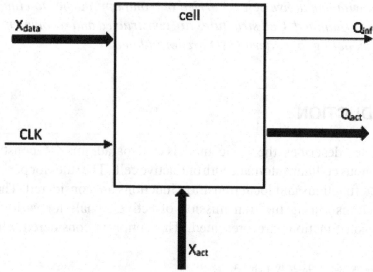

AC is characterized by the fact that it forms an information signal (a state signal at the current moment of time) and an additional signal of the active state. The signal of the active state indicates that at the next moment in time one of the cells in the neighborhood will go into the active state. The generalized cell structure of such a cellular automaton on Figure 1 is shown.

The cell contains information inputs (X_{data}), which are connected to the information outputs (Q_{inf}) of all cells in the neighborhood, inputs of active states (X_{act}), which are connected to the corresponding outputs of the active states (Q_{act}) of the cells. Each cell also contains (but may not contain) a synchronization input (CLK).

Depending on the active state transmission mode, the number of active outputs may vary. If a mode is used in which the active cell itself selects the next active cell from the cells in the neighborhood, then the number of active outputs is greater than 1. In this case, the cell contains n active inputs $X_{act,1}$,..., $X_{act,n}$ (were n – the number of neighborhood cells), the number of which is equal to the number of cells in the neighborhood. In the case when each cell in the neighborhood of the active cell itself determines its next state of activity, then the active output of Q_{act} can be one. In this case, the cell contains n active inputs $X_{act,1}$,..., $X_{act,n}$.

According to the presented graphical interface of the cell (Figure 1), the following models can describe the cell.

$$Q_{inf}(t+1) = \begin{cases} f_{state}\left[X_{data,1}(t),...,X_{data,n}(t)\right], if\ X_{act}(t) = 1 \\ Q_{inf}(t), if\ X_{act}(t) = 0 \end{cases}, \quad (4)$$

$$Q_{inf}(t+1) = \begin{cases} f_{state}\left[X_{data,1}(t),...,X_{data,n}(t)\right], if\ f_{LTF}\left[X_{act,1}(t),...,X_{act,n}(t)\right] = 1 \\ Q_{inf}(t), if\ f_{LTF}\left[X_{act,1}(t),...,X_{act,n}(t)\right] = 1 \end{cases}, \quad (5)$$

were $f_{state}\left[X_{data,1}(t),...,X_{data,n}(t)\right]$ – local state function of cell.

Model (4) implements the first active state transmission mode, and model (5) implements the second active state transmission mode. An example of the functioning of an ACA with one active cell on Figure 2 is shown.

However, there are possible modes in which the transfer of the active state can be carried out only between two neighboring cells during the entire time

Figure 2. An example of the functioning of an ACA with one active cell

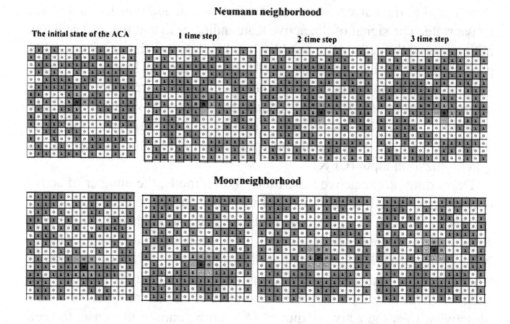

of functioning of the cellular automaton. Such a cycle for this example (Figure 2) is presented at the third time step for the von Neumann neighborhood. Examples of such situations on Figure 3 are presented.

Figure 3. Examples of ACA with one active cell with single cycles

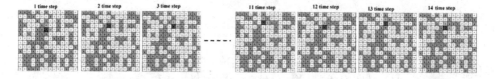

The cycle starts at time step 11 and the ACA functioning continues the rest of the time.

In addition to single-cycle cycles, there are cycles that, after a longer period of time, return to their one of the previous states. The more time ACA works until it comes to one of its initial states, the longer its life cycle. Examples of arising cycles in the work of ACA with one active cell on Figure 4 are presented.

Figure 4. Examples of cycles in the work of ACA with one active cell

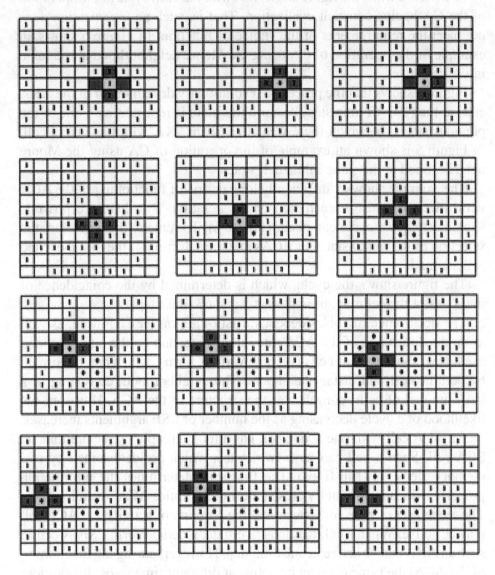

This example shows the process of moving the excitation signal on the CA cells. The figure also shows how the state of each excited cell changes at the appropriate time. The state of the excited cells changes according to the expression

$$a_{\bullet\bullet}(t+1) = a_0(t) \oplus a_1(t) \oplus a_2(t) \oplus a_3(t).$$

At the same time, the figure shows that with the 10th cycle of the operation of the CA begins constant cyclization. That is, such an organization does not meet the requirements of the choice of function. In Figure 5 shows an example of the operation of CA using the Moore neighborhood at the same initial cell states.

Figure 4 shows that the process of evolutionary development of the AKA ends at the (10)-th time step, because this state is repeated with one of the previous states, in which there was already an AKA with one active cell. In Figure 5 is shown an example of the operation of CA using the Moore neighborhood at the same initial cell states

The example shows a sufficiently long period of functioning of the ACA with different states at each step of the iteration. But there are possible situations when a cycle can occur (Figure 6). To remedy this situation, the selection of the initial states of a CAs and the selection of the initial excited cell are made.

The figure shows the cycle, which is determined by the coincidence of the states of the cellular automaton at 3 and 63 time steps.

The occurrence of cycles depends, first of all, on the LTF used, as well as on the dimension of the ACA, the initial state, and on the neighborhood used. Neighborhoods of cells can take various forms, which can be formed by both nearest and distant neighbors. The neighborhood, consisting of a large number of cells, complicates the structure of the ACA. However, the likelihood of a cycle decreasing as the number of LSF arguments increases.

An ACA cell with one active cell has a more complex structure than an SCA cell. Such a cell can have two states, which are determined by two memory elements (flip-flops). The first state determines the information property, which is constantly present during the entire ACA operation time. The second state of the cell determines the property of activity, i.e. ACA cell may be active. When a cell has an active state, it can perform LSF. A cell in such an AKA can be active in one time step. However, during the functioning of the ACA, the same cell can be active at different time steps. In addition, the cell may not have an active state during the entire functioning of the ACA.

At each time step, the state of the ACA may differ from the state of the ACA at the previous time step only in the position of one cell, but it may not differ. The difference lies in the performance of the LSF by an active cell. Significantly change the state of the ACA is possible only if the ACA will work a large number of cycles and almost all cells of the ACA will be active in this period of time.

Figure 5. An example of the operation of a CA using the Moore neighborhood

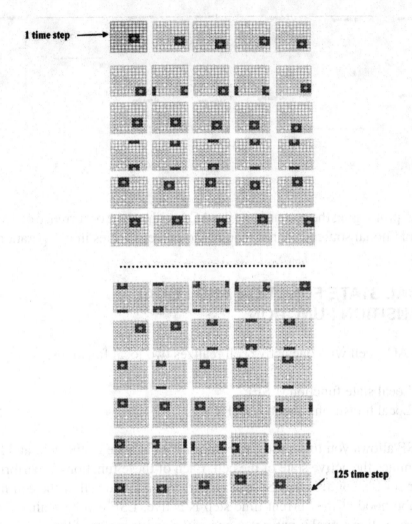

To reduce the time of functioning of the ACA for a substantial and rapid change in the state of the ACA, as well as to eliminate LSF cycles, cells in the neighborhood of the active cell can perform. In this case, one of the cells in the neighborhood becomes active according to one of two options. An example of the functioning of such an ACA with one active cell on Figure 7 is shown.

Obviously, the states of the ACA at neighboring time steps are different from each other. The active signal is transmitted to the cell of the neighborhood after the LSF by all cells of the neighborhood. Such organization of the ACA

Figure 6. An example of a cycle in a CA based on the von Neumann neighborhood

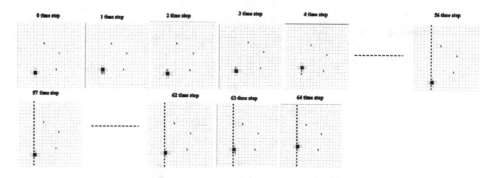

allows prolonging the operation of the ACA without the occurrence of cycles, and additional sources are not required to prevent cycles in its operation.

LOCAL STATE FUNCTION AND LOCAL TRANSITION FUNCTION

Each ACA cell with one active cell realizes two local functions:

- Local state function (LSF);
- Local transition function (LTF).

LSF allows you to set the basic informational state of the cell, and LTF determines the active state of the cell. Each of these functions is performed under certain conditions. LSF can be performed if the cell or the cell in its neighborhood at the current time step is active. LSF can be realized only when an active signal is present at one of the active inputs of the cell. Those. a cell can perform LFP if one of its active cells is active.

Figure 7. An example of the functioning of ACA with one active cell and the implementation of LSF cells in the neighborhood

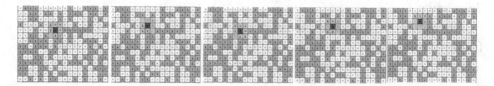

However, each active cell has several neighboring cells at the current time step. This does not mean that all of these neighborhood cells performs LTF. In this situation, the implementation of LTF depends on the variant of internal implementation of the ACA cell.

There are several options for the implementation of cells that perform LSF at the current time step.

1. The cell is active.
2. One of the cells in the neighborhood is active.
3. The cell and its cells of neighborhood are not active, but the internal LSF of the cell has a logical value of "1".

The first option is described by a model of the form

$$LSF(t+1) = \begin{cases} f\left[X_1(t),...,X_n(t)\right], if\, Y_{act}(t) = 1 \\ S(t-k), if\, Y_{act}(t) = 0 \end{cases}, \qquad (6)$$

where

Y_{act} – signal of the active state at the active output of the cell;

$X_i(t)$ – information signal at the information input of the ith cell of the neighborhood $\left(i = \overline{1,n}\right)$;

S(t-k) – result of LSF execution by a cell at an early (t-k) th time step.

S(t-k) determines the time step of ACA functioning, on which the cell was active the last time before the current time step. If the cell has never been active, then its information state remains the initial state (S_0). Thus, in the first option of implementation, the cell monitors its own state of activity.

In addition, in this mode, the cell in the first mode performs LTF, which is described by the following model

$$LTF_{cell}(t+1) = \begin{cases} f_{act}\left[X_1(t),...,X_n(t)\right], if\, \vee_n^{i=1} X_{i,act}(t) = 1 \\ 0, if\, \vee_n^{i=1} X_{i,act}(t) = 0 \end{cases}, \qquad (7)$$

$$f_{act}\Big[X_1(t),\dots,X_n(t)\Big] = \Big\{Y_{i,act}(t+1)\Big\} = \sum_{i=1}^{n} y_i(t+1) = 1,$$

where

$X_{i,act}(t)$ – signal at the i-th active input of the cell at time step t;

n – number of neighborhood cells;

$y_i(t+1)$ – signal at the i-th active cell output at time (t+1).

Model (2) implements a logical function that analyzes signals from information outputs (function arguments) and generates a result on n active outputs. In this case, a logical "1" signal is generated at only one output $Y_{i,act}(t+1)$. This output is connected to the active output of the neighborhood cell, which in the next time step will become active and will perform LSF.

For the second option, LFS is described by the following model

$$LSF(t+1) = \begin{cases} f_1\Big[X_1(t),\dots,X_n(t)\Big], \text{ if } f_2\Big[X_1(t),\dots,X_n(t)\Big]f_{act}\Big[X_{1,act}(t),\dots,X_{n,act}(t)\Big] = 1 \\ S(t-k), \text{ in other case} \end{cases},$$

$$(8)$$

where

$f_1\Big[X_1(t),\dots,X_n(t)\Big]$ – local state function that establishes the basic informational state of the cell;

$f_2\Big[X_1(t),\dots,X_n(t)\Big]$ - local state function that allows the cell to go into an active state;

$f_{act}\Big[X_{1,act}(t),\dots,X_{n,act}(t)\Big]$ – local transition function, the arguments of which are signals at the active outputs of the neighborhood.

The second option involves the transition of the cell into an active state using two functions $f_2\Big[X_1(t),\dots,X_n(t)\Big]$ and $f_{act}\Big[X_{1,act}(t),\dots,X_{n,act}(t)\Big]$. These two functions actually form the general active function of the cell, which is represented by the following mathematical model

$$LTF_{cell}\left(t+1\right)=\begin{cases} f_{act,cell}\left[X_{1}\left(t\right),...,X_{n}\left(t\right),X_{1,act}\left(t\right),...,X_{n,act}\left(t\right)\right] \\ = f_{2}\left[X_{1}\left(t\right),...,X_{n}\left(t\right)\right]f_{act}\left[X_{1,act}\left(t\right),...,X_{n,act}\left(t\right)\right], \\ 0, in\ other\ case \end{cases} \quad (9)$$

where $f_{act,cell}\left[X_{1}\left(t\right),...,X_{n}\left(t\right),X_{1,act}\left(t\right),...,X_{n,act}\left(t\right)\right]$ – general active function that puts a cell in an active state.

The third option uses the LSF mathematical model (6). LSF is performed only in the case of its transition to an active state. However, for the third option, the cell becomes active regardless of the presence of neighboring active cells. The cell becomes active as a result of the analysis of the informational states of the neighborhood cells. Cells in an active state do not transmit an active state to neighboring cells in the neighborhood. LTF is described by the following mathematical model

$$LTF\left(t+1\right)=\begin{cases} Y_{act}\left(t\right)=1,if\ f_{act}\left[X_{1}\left(t\right),...,X_{n}\left(t\right)\right]=1 \\ 0, in\ other\ case \end{cases} \quad (10)$$

The model shows that it does not contain active signals from neighboring cells. This situation can lead to the appearance of several active cells in the next time step, and can lead to their complete disappearance. Therefore, the third option is not always acceptable for constructing an ACA with one active cell.

The described models show that LSF and LTF are interdependent. A cell can become active in the next time step if at the current time step one of the cells in its neighborhood is active. LSF is performed if the LTF of the same cell has a logical value of "1" (the cell is active). Also, the cell becomes active in the next time step if the combination of states of the neighborhood cells gives $LTF(t+1)=1$.

MODEL OF CELLULAR AUTOMATA WITH CONTROLLED MOVEMENT OF THE ACTIVE CELLS

The models and options for implementing ACA with one active cell, considered in the previous paragraphs, carry out pseudo-controlled movement of the active state along ACA cells. The trajectory of the active state depends on the initial settings of the ACA, the choice of the initial active cell, LSF and LTF.

If the initial settings of the ACA are known, then it is very difficult to determine the location of the active cell at the selected time step. To know the trajectory of the active state, it is necessary to initially determine and describe the functioning of the ACA at the selected initial settings.

However, the larger the dimension of the ACA, the larger the initial settings can be, which are determined by the number of states of the logical "1" and "0". For example, if the dimension of the ACA corresponds to 10×10, then the number of possible states is 2^{100}. It is difficult to trace such a number of initial conditions. Therefore, not in all cases, it is possible to predict the trajectory of the active state in the cells of the ACA.

In modern systems of intelligent processing and information protection, there are tasks in which there is a need to set the trajectory of the active signal. For example, such tasks include image processing tasks (description of the image contour, handwriting recognition, etc.). In such problems, controlled movements of the active state along cells of a certain property are used. For example, an active signal can propagate through the cells of a certain state of the neighborhood cells. An example of the movement of the active state through the cells of the ACA when tracking the trajectory of writing a handwritten letter is presented in Figure 8.

Figure 8. Examples of moving an active state through the cells of a handwritten symbol

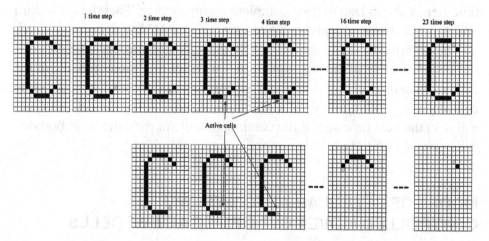

In the presented example, a binary image of the handwritten symbol is formed in time, and the active state moves along the cells belonging to the handwritten symbol. In the first variant of distribution, the previous active cell does not go into the state of logical "1". In the second embodiment, in addition to activity, the cell, being in the active state in the current time step, sets the state in the logical "0" state to the cell in the active state in the previous time step. Active signal propagates only in cells having a logical state of "1". In this mode, the AKA controls the movement of the active signal through cells having the logical state property of "1".

To implement the motion control mode of the active signal, the following mathematical model is used

$$S(t+1) = \begin{cases} LSF_1(t), if \vee_n^{i=1} X_{i,act}(t) = 1, LSF_0(t) = 1 \\ S(t), if \vee_n^{i=1} X_{i,act}(t) = 0, Y_{act}(t) = 0 \\ LSF_2(t), if \vee_n^{i=1} X_{i,act}(t) = 0, Y_{act}(t) = 1 \end{cases}, \tag{11}$$

where

LSF_0 – local state function that determines the transition of a cell to an active state;

LSF_1 – local state function, which is performed if one of the neighboring cells is active and controlled by $LSF_0 = 1$;

LSF_2 – local state function, which is performed after the transfer of an active state to a neighboring cell.

The control function is LSF_0. In fact, this function determines the direction of movement. For the example presented on Figure 8 the LSF_0 is described by the following expression

$$LSF_0 == (X_1 \overline{X_2 X_3 X_4 X_5 X_6 X_7 X_8})(\overline{X_1} X_2 \overline{X_3 X_4 X_5 X_6 X_7 X_8})(\overline{X_1 X_2} X_3 \overline{X_4 X_5 X_6 X_7 X_8})$$
$$(\overline{X_1 X_2 X_3} X_4 \overline{X_5 X_6 X_7 X_8})(\overline{X_1 X_2 X_3 X_4} X_5 \overline{X_6 X_7 X_8})(\overline{X_1 X_2 X_3 X_4 X_5} X_6 \overline{X_7 X_8})$$
$$\vee \overline{X_1 X_2 X_3 X_4 X_5 X_6} X_7 \overline{X_8})(\overline{X_1 X_2 X_3 X_4 X_5 X_6 X_7} X_8)$$

LSF_0 in this example is described for the Moore neighborhood and determines the presence of one cell of the neighborhood in a state of logical "1".

If the cell changes its state and properties after the transfer of the active state, then LSF_2 is used. A cell changes its state or property.

The cell of such an ACA is represented by the following interface structure in the form of a black box (Figure 9).

Figure 9. ACA cell interface with controlled movement of the active signal

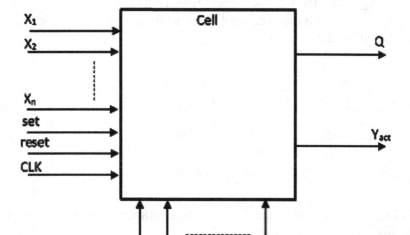

The internal structure of such a cell on Figure 10 is shown.

The cell contains three combinational circuits (CC_0, CC_1, CC_2), which respectively implement LSF_0, LSF_1 and LSF_2. The first flip-flop (T_1) determines the informational state of the cell, and the second flip-flop (T_2) determines the active state of the cell. The second trigger T_2 enters the logical "1" state ($Y_{act} = 1$) if a logical "1" signal is present at the CC_0 output and a logical "1" signal is also present at one of the active inputs ($X_{i,act}$) of the cell.

The active state of the cell ($Y_{act} = 1$) controls the state of the first flip-flop. On the leading edge of the signal at the output Y_{act}, the first flip-flop (T_1) takes a state in accordance with the operation CC_1 ($Q=LSF_1$). At a given edge of the active signal, the first flip-flop (T_1) takes a state according to CC_2 ($Q=LSF_2$). The Reset input on the second flip-flop resets the active signal at the next time. Logical signal "1" comes from the active output of one of the neighboring cells and puts the second flip-flop in the state of logic "0" on the leading edge with a given delay.

Figure 10. Functional diagram of an ACA cell with controlled movement of the active signal

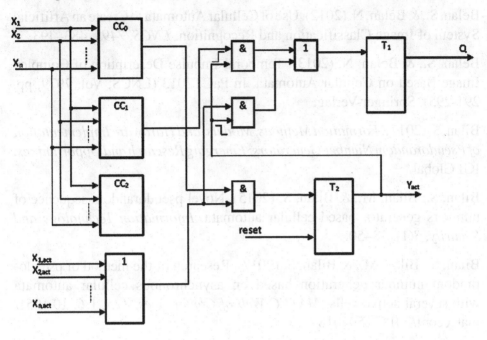

This organization of the cell structure uses additional delay elements and memory elements. Cell modeling was carried out and its VHDL model was obtained.

CONCLUSION

The proposed CA with one active cell made it possible to formulate the basic principles of the functioning of AKA with active cells that function with two states. The informational state of the cell is inherent in previous CAs, and the active state describes the process of transferring the active state from cell to cell. An additional state is used to process various CA states and to move active cells in a given direction in the CA field. This approach has a number of advantages that allow us to describe many dynamic processes that could not be described by classical CA.

REFERENCES

Belan, S., & Belan, N. (2012). Use of Cellular Automata to Create an Artificial System of Image Classification and Recognition. *LNCS, 7495*, 483–493.

Belan, S., & Belan, N. (2013). Temporal-Impulse Description of Complex Image Based on Cellular Automata. In PaCT2013 (LNCS, Vol. 7979, pp. 291-295). Springer-Verlag.

Bilan, S. (2017). *Formation Methods, Models, and Hardware Implementation of Pseudorandom Number Generators: Emerging Research and Opportunities*. IGI Global.

Bilan, S., Bilan, M., & Bilan, S. (2015). Novel pseudorandom sequence of numbers generator based cellular automata. *Information Technology and Security, 3*(1), 38–50.

Bilan, S., Bilan, M., & Bilan, S. (2017). Research of the method of pseudo-random number generation based on asynchronous cellular automata with several active cells. *MATEC Web of Conferences, 125*, 1-6. 10.1051/matecconf/201712502018

Bilan, S., Bilan, M., Motornyuk, R., Bilan, A., & Bilan, S. (2016). Research and Analysis of the Pseudorandom Number Generators Implemented on Cellular Automata. *WSEAS Transactions on Systems, 15*, 275–281.

Chapter 4
Models and Paradigms of Cellular Automata With Several Active Cells

ABSTRACT

The chapter describes the models and paradigms of asynchronous cellular automata with several active cells. Variants of active states are considered in which an asynchronous cellular automaton functions without loss of active cells. Structures that allow the coincidence of several active states in one cell of a cellular automaton are presented. The cell scheme is complicated by adding several active triggers and state control schemes for active triggers. The VHDL models of such cells were developed. Attention is paid to the choice of local state functions and local transition functions. The local transition functions are different for each active state. This allows you to transmit active signals in different directions. At each time step, two cells can change their information state according to the local state function. Asynchronous cellular automata have a long lifecycle.

INTRODUCTION

The chapter describes the models and paradigms of asynchronous cellular automata with several active cells. Variants of active states are considered in which an asynchronous cellular automaton functions without loss of active cells. Structures that allow the coincidence of several active states in one cell

DOI: 10.4018/978-1-7998-2649-1.ch004

of a cellular automaton are presented. The cell scheme is complicated by adding several active triggers and state control schemes for active triggers. The VHDL - models of such cells was developed. Attention is paid to the choice of local state functions and local transition functions. The local transition functions are different for each active state. This allows you to transmit active signals in different directions. At each time step, two cells can change their information state according to the local state function. Asynchronous cellular automata have a long life cycle.

MODEL AND PRINCIPLES OF FUNCTIONING OF CELLULAR AUTOMATA WITH SEVERAL ACTIVE CELLS

Based on the ACA with one active cell, pseudorandom number generators (PRNGs) were constructed, which were investigated using statistical tests ENT, NIST (Bilan, 2017; Bilan, Bilan, & Bilan, 2015; Bilan, et al 2016), as well as using graphical tests (Bilan, 2017). Studies have allowed the analysis of ACA with one active cell. However, ACA studies with several active cells have not been conducted. Studies have been conducted for an ACA with one active cell and for various forms of organizing neighborhoods. In such ACAs, at each time step, only one cell changes its state.

Studies show that the small dimension of an ACA with one active cell does not give a long life cycle. The increase in the number of active cells allows you to extend the length of the ACA life period.

The structure of an ACA with several active cells has the same structure as an ACA with one active cell. Moreover, each ACA cell with several active cells has more advanced functional capabilities. A cell of such an ACA has a different structure since it can operate in four modes. It is taken into account that ACA contains two active cells.

- Standby mode;
- The mode of the first active cell;
- The mode of the second active cell;
- The mode of the first and second active cells.

The standby mode is characterized by the fact that the cell is not active and is in a logical state of "1" or "0". In the active state, the cell performs LSF and can change its information state according to LSF. The result of

LSF execution depends on the values of the arguments that are formed at the outputs of the cells of its neighborhood. The argument can also be the value at the output of the most active cell in the previous time step. The second mode refers to the active cell. The second mode is similar to the active cell mode for an ACA with one active cell, which was discussed in the previous section.

The mode of the second active cell is similar to the mode of the first active cell. However, the difference is that the second active cell performs another LTF.

The fourth mode corresponds to a cell that is in the states of the first and second active cells at the current time step. In this mode, the active cell cannot simultaneously execute two different LSFs, so each cell has one information state (one information output). In such a situation, the following submodes may occur:

- LSF with high priority;
- Each of the LSFs is performed at a specific time step;
- An LSF is executed, the argument of which is the resulting signals of the first and second LSF.

In the first submode of one of the active cells, a higher priority is given. The LSF of this cell is always performed for the fourth cell mode. In the second submode, both active cells perform the same LSF but at different time steps (for example, the odd time steps implement the LSF of the first active cell, and the second active cell performs LSF at the even time steps) (Bilan, 2017; Bilan, Bilan, & Bilan, 2015; Bilan, et al 2016).

If the active cell is in the third submode, it performs a common LSF. For example,

$$LSF_{\Sigma} = LSF_1 \oplus LSF_2$$

Total LSF_{Σ} can implement any logical function.

According to the described modes, the cell can be represented as a black box model (Figure 1).

Given all the possible modes of the ACA cell, its behavior can be described by the following model

Figure 1. ACA cell black box model with multiple active cells

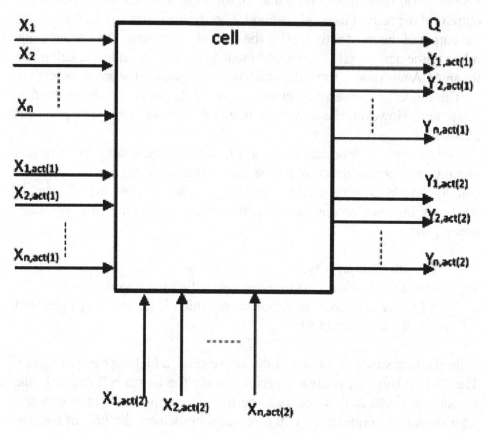

$$S(t+1) = \begin{cases} f_1\Big[x_1(t),\dots,x_n(t)\Big], if \vee_{i=1}^{n} x_{i,act(1)} = 1, \vee_{i=1}^{n} x_{i,act(2)} = 0 \\ f_2\Big[x_1(t),\dots,x_n(t)\Big], if \vee_{i=1}^{n} x_{i,act(1)} = 0, \vee_{i=1}^{n} x_{i,act(2)} = 1, \\ f_1\Big[x_1(t),\dots,x_n(t)\Big], if \vee_{i=1}^{n} x_{i,act(1)} = \vee_{i=1}^{n} x_{i,act(2)} = 1 \end{cases} \qquad (12)$$

where

$x_{i,act(1)}$ – the signal at the ith active input of the first group, which is formed at the active outputs of $Y_{i,act(1)}$ cells in the neighborhood and determines the first active state of the cell;

$x_{i,act(2)}$ – a signal at the ith active input of the second group, which is formed at the active outputs; $Y_{i,act(2)}$ - a signal at the outputs of active neighborhood cells, which determines the second active state;

$f_1\left[x_1(t),...,x_n(t)\right]$ – LSF, which is executed if the cell goes into the first active state;

$f_2\left[x_1(t),...,x_n(t)\right]$ - LSF, which is performed when a cell goes into a second active state;

$f_3\left[x_1(t),...,x_n(t)\right]$ - LSF, which is performed if the cell enters the first and second active states at the same time step;

Local state functions $f_1\left[x_1(t),...,x_n(t)\right]$, $f_2\left[x_1(t),...,x_n(t)\right]$ and $f_3\left[x_1(t),...,x_n(t)\right]$ may be equal, but may be different. It depends on the AKA problem being solved. A cell can execute different LSFs at two adjacent time steps. Also, the cell may not go into an active state during the entire functioning of the ACA, and the cell may have only one in two active states.

Due to the fact that the number of active states increases, the number of active inputs also increases. The transition of the cell into an active state can be described by the following model.

$$S_{act}(t+1) = \begin{cases} S_{act(1)}, if\ f_{act(1)}\left[x_{1,act(1)},...,x_{n,act(1)}\right] = 1, f_{act(2)}\left[x_{1,act(2)},...,x_{n,act(2)}\right] = 0 \\ S_{act(2)}, if\ f_{act(1)}\left[x_{1,act(1)},...,x_{n,act(1)}\right] = 0, f_{act(2)}\left[x_{1,act(2)},...,x_{n,act(2)}\right] = 1 \\ S_{act(1,2)}, if\ f_{act(1)}\left[x_{1,act(1)},...,x_{n,act(1)}\right] = f_{act(2)}\left[x_{1,act(2)},...,x_{n,act(2)}\right] = 1 \\ 0, in\ other\ case \end{cases}$$

(13)

where

$S_{act(i)}$ – i-th active state of the cell;

$f_{act(i)}\left[x_{1,act(i)},...,x_{n,act(i)}\right]$ – i-th LTF.

However, the cell can go into the active state taking into account the fulfillment of the given LSF. This approach reduces the number of active inputs and is described by the following model.

$$S_{act}\left(t+1\right) = \begin{cases} S_{act(1)}, if\ f_1\left[x_1\left(t\right),\ldots,x_n\left(t\right)\right] = 1, \vee_{i=1}^n x_{i,act} = 1, f_2\left[x_1\left(t\right),\ldots,x_n\left(t\right)\right] = 0 \\ S_{act(2)}, if\ f_1\left[x_1\left(t\right),\ldots,x_n\left(t\right)\right] = 0, \vee_{i=1}^n x_{i,act} = 1, f_2\left[x_1\left(t\right),\ldots,x_n\left(t\right)\right] = 1 \\ S_{act(1,2)}, if\ f_1\left[x_1\left(t\right),\ldots,x_n\left(t\right)\right] = f_2\left[x_1\left(t\right),\ldots,x_n\left(t\right)\right] = 1, \vee_{i=1}^n x_{i,act} = 1, \end{cases}$$

$$(14)$$

The model shows that a cell can go into one of the active states in the current time step, provided that at the previous time step one of the cells in the neighborhood was active. In this case, the cell can be in one of the active states, and in the next time step, one of the cells in the neighborhood can go into another active state and this cell will perform another LSF. In such a situation, at a certain point in time, there can be several active cells in the CA field with the same active state or different. A situation may also arise when there will be one active cell in the CA field. This is due to fewer active outputs and inputs. To eliminate possible switching of active states, the number of active inputs and outputs must correspond to each active state. A situation may arise when LSF cells in the neighborhood of the active cell are not equal to the logical "1". Then the active cell is in standby mode until the cells in the neighborhood change their state, which will allow them to get a local function equal to the logical "1".

A change in the state of cells can occur as a result of a transition to another active state at a certain time step. The number of active cells may be more than two. Then the CA cell structure becomes more complex.

ALGORITHM OF CELLULAR FUNCTIONING WITH SEVERAL LOCAL TRANSITION FUNCTION

Unlike an ACA with one active cell, an ACA cell with several active and cells has a more complex structure. This is due to the fact that each active state is determined by a separate memory element (flip-flop). Each such flip-flop determines an active state while in a logical "1" state. If cells with two active states are used in the ACA, then the cell has inputs and outputs shown in Figure 1.

The algorithm of the cell is as follows.

1. The cell in the information state (logical "0" or "1") is installed.
2. The presence of active signals at the inputs of the initial installation of the cell in an active state is determine.
    2.1. If active signals are not present at the active inputs, then the cell goes into standby mode (go to step 2).
    2.2. If the active signal (logical signal "1") is present at one or more active inputs, then the cell goes into an active state. Go to step 3.
3. The corresponding LSF of the cell is performed and the cell enters the information state according to the LSF.
4. LTF cells are performed for all active states.
5. The formation of the active signal at the corresponding active inputs of the cell.
6. Zeroing the active state of the cell.
7. Analysis of the ACA shutdown signal.
    7.1. If the end signal is "0", then go to step 2.
    7.2. If the end signal is "1", then go to step 8.
8. The end.

The generalized verbal algorithm is described for a finite number of time steps of the ACA. For this, a shutdown signal is used. However, the cell can work according to this algorithm an infinite number of time steps. If a small amount of active states is used, then the cell is in standby mode for an active signal from cells in the neighborhood for more time. If a large number of active states are used, the time taken to wait for active signals is reduced. Also affects the dimension of the ACA and the shape of the neighborhood. The more cells form a neighborhood, the greater the likelihood of a cell transitioning to an active state at the next time step.

The graph - a diagram of the algorithm of the cell with several active states on Figure 2 is presented.

The algorithm is described for cell operation in one time step of ACA operation. The cell may be in standby mode for a long time. Standby ends with the arrival of an active signal at one of the active inputs.

The algorithm also describes a cell with two active states. If to increase the number of active states of the cell, then conditional and operator vertices for each active state are added accordingly (4 conditional vertices and 2 operator vertices). Such an algorithm cannot be effectively implemented programmatically on a single-processor computer, since it has a sequential character of enumerating all active signals at active inputs. However, the beginning of the algorithm can be parallelized and, after the first operator vertex,

61

Figure 2. The graph - a diagram of the algorithm of the cell with several active states

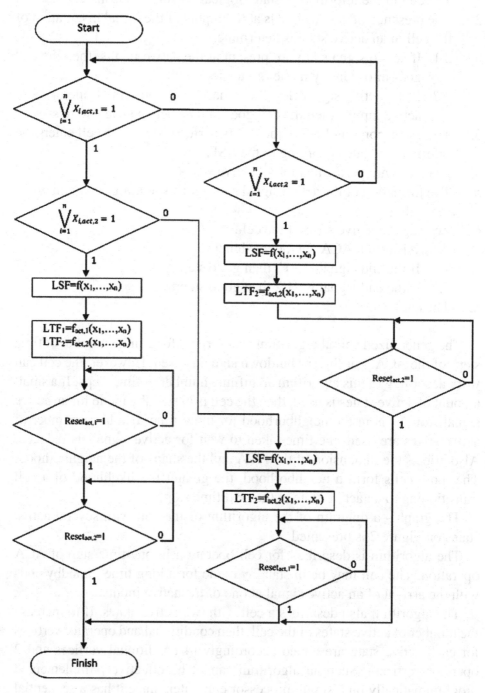

simultaneously execute conditional algorithm blocks. Such parallelization requires additional hardware resources and memory costs. The program code is also complicated. Therefore, such an ACA is better to implement hardware.

Given the greater number of active states (> 2), the transition of a cell to one of the active states is determined by the following section of the graph diagram of the algorithm (Figure 3).

Figure 3. Graph diagram of the algorithm of the cell for n active states

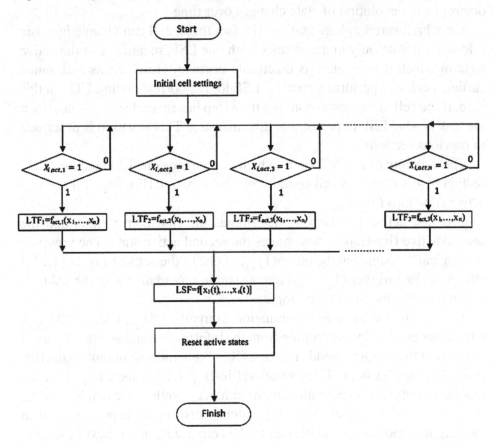

The algorithm presented in Figure 3 does not describe resetting active states. Moreover, the process of zeroing active states also has a parallel structure. This algorithm can be implemented on a multiprocessor structure and requires a lot of time in software implementation.

HARDWARE IMPLEMENTATION OF CELLS WITH MULTIPLE LOCAL TRANSITION FUNCTIONS

ACA implemented on cells that can have several active states requires a lot of time to implement the algorithm of one cell. Software implementation does not allow parallel work of all ACA cells. Therefore, the path of hardware implementation of the ACA was chosen. Software implementation is used to simulate the evolution of the development of ACA. The resulting software model allows us to evaluate the organization of the ACA, as well as to determine the evolution of state changes over time.

For a hardware implementation, the fact that a cell can change its basic information state only in accordance with one LSF, regardless of the active state in which it is located, is taken into account. However, as mentioned earlier, a cell can perform a specific LSF for the corresponding LTF. In this case, if the cell at the corresponding time step has several active states, then the task is what LSF to perform at this time step. This situation is described in previous sections.

The structure of the ACA cell, which implements one LSF and two LTF, on Figure 4 is shown. A cell contains one information flip-flop (T_1) and two active flip-flops (T_2, T_3).

The first active flip-flop T_2 determines the first active state, and the second active flip-flop T_3 determines the second active state. The presence of a logical "1" signal at the output (y_{act}) of one of the active triggers (T_2, T_3) allows the first trigger (T_1) to go into a state in accordance with the value of signal 0 or 1 at its information input.

The circuit also contains combinational circuits (CU_{LSF}, CU_{LTF1}, CU_{LTF2}), which process the signals coming from the information and active outputs of the cells of the neighborhood. The first CU_{LSF} generates an output signal that controls the setting of the T_1 information flip-flop. CU_{LTF1} and CU_{LTF2} have the number of outputs corresponding to the number of cells in the neighborhood. A single signal at one of the CU_{LTF} outputs sets the corresponding cell in the neighborhood of the active cell to the active state at the next time step. CU_{LTF1} and CU_{LTF2} control the flip-flop T_2 and T_3. If T_2 and (or) T_3 are in a single state, then the second and third combinational circuits implement the corresponding LTF.

The second and third triggers T_2, T_3 go into the logical "1" state (active state), if the logical "1" signal is present at the outputs of the multi-input OR

Figure 4. Functional diagram of an ACA cell with two active cells

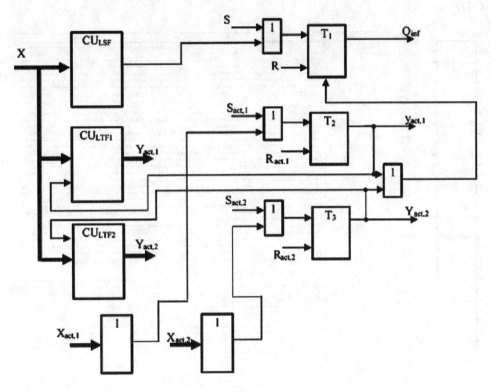

- gate. The inputs of these elements receive signals from the active outputs of neighboring cells.

The circuit of such a cell is modeled in Active-HDL CAD (Figure 5) and can be implemented on FPGA.

For simulation, the synchronized flip-flops were used. Testing of this scheme showed high reliability of a cell functioning. In the implementation considered, only one LSF is executed for any active state. However, another embodiment of the cell is also possible when a separate LSF is performed for each active state. A fragment of the functional diagram for such an implementation option in Figure 6 is shown. The figure shows a part of the ACA cell scheme, which implements the function of controlling the basic (informational) state of the cell.

The first and second combinational circuits (CU_{LSF1}, CU_{LSF2}) execute user-selected LSFs, and the third combinational circuit (CU_{LSF3}) implements a logical function of two arguments, which are signals from the outputs of the

Figure 5. Functional modeling of an ACA cell with two active cells in Active-HDL CAD

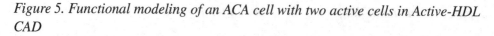
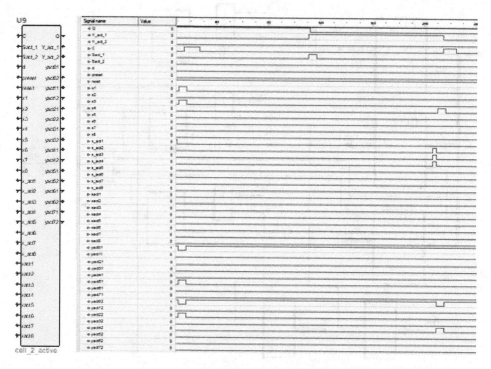

first and second combinational circuits (CU_{LSF1}, CU_{LSF2}). All combinational circuits are controlled by signals from the outputs of active flip-flop.

With an increase in active states, the number of combinational circuits implementing LTF is increases. The number of combinational circuits implementing LSF for the second option is also increasing.

The simplicity of implementation and modeling of AKA cell schemes with several active cells allows you to create homogeneous schemes based on the hierarchical implementation of projects in modern CAD systems for FPGAs of leading companies such as Xilinx, Altera and other.

Programs have also been developed that simulate the functioning of the ACA with two active cells. Modeling was performed for orthogonal and hexagonal coverage ACA. Each active cell is indicated by a separate color. Examples of the functioning of the ACA with two active cells on Figure 7 are presented. Examples are presented for two forms of coverage of a cellular automaton (orthogonal and hexagonal).

Figure 6. A fragment of the information part of an AKA cell with several LSFs

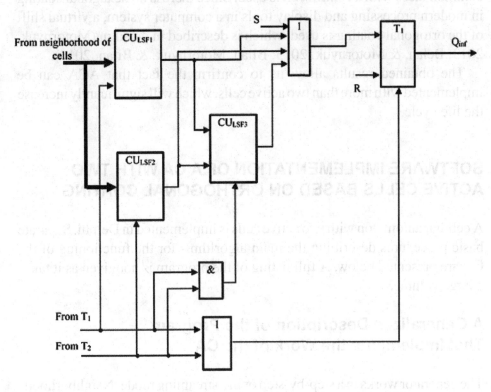

During one time step, two cells that are active at this point in time change their state. For an orthogonal coated ACA, the Moore neighborhood is used, and for a hexagonal coated ACA, a neighborhood consisting of six cells sharing

Figure 7. An example of the functioning of an ACA with two active cells for orthogonal and hexagonal coating

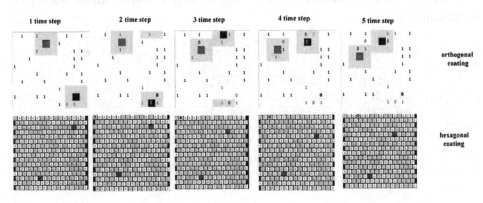

common faces with a control cell is used. Since there is no hexagonal coating in modern processing and display tools in a computer system, a virtual shift of the ortogonal coating is used, which is described in detail in (Motornyuk, 2013; Belan, & Motornyuk. 2013; Bilan, Motornyuk, & Bilan, 2014).

The obtained results allow us to confirm the fact that ACA can be implemented with more than two active cells, which will significantly increase the life cycle.

SOFTWARE IMPLEMENTATION OF A CA WITH TWO ACTIVE CELLS BASED ON ORTHOGONAL COATING

A cellular automaton with two active cells is implemented in Delphi. Separate basic procedures describing the main algorithms for the functioning of the CA are presented below. A full listing of the program is not given as it takes a large volume.

A Generalized Description of the Program That Implements the Work of the CA

The generator works in a step-by-step or in a streaming mode. Neighborhoods are processed in two types - according to Moore or according to Neumann. An array of extra bits is formed. If the flag "after each step" is set, then an array of additional bits is formed after each step. If the flag "after (n $\times$ m)-th step" is set, then an array of additional bits is formed after reaching the end of the workpiece.

The calculation of the new value of the next cell is carried out by XOR operation of the current cell with all cells in the neighborhood and an additional bit, which is taken from the array of additional bits. Listing snippet is presented below.

```
Sled:= MasC[XYCur.X,XYCur.Y].value xor MasC[Nxy[1].X,Nxy[1].Y].
value xor
        MasC[Nxy[2].X,Nxy[2].Y].value xor
MasC[Nxy[3].X,Nxy[3].Y].value xor
        MasC[Nxy[4].X,Nxy[4].Y].value xor
MasC[Nxy[5].X,Nxy[5].Y].value xor
        MasC[Nxy[6].X,Nxy[6].Y].value xor
MasC[Nxy[7].X,Nxy[7].Y].value xor
```

```
            MasC[Nxy[8].X,Nxy[8].Y].value xor BitD;
      MasC[XYCur.X,XYCur.Y].value:= Sled;
```

The value of the second control point is calculated in the same way.

```
Sled2:= MasC[XYCur2.X,XYCur2.Y].value xor
MasC[Nxy2[1].X,Nxy2[1].Y].value xor
         MasC[Nxy2[2].X,Nxy2[2].Y].value xor
MasC[Nxy2[3].X,Nxy2[3].Y].value xor
         MasC[Nxy2[4].X,Nxy2[4].Y].value xor
MasC[Nxy2[5].X,Nxy2[5].Y].value xor
         MasC[Nxy2[6].X,Nxy2[6].Y].value xor
MasC[Nxy2[7].X,Nxy2[7].Y].value xor
         MasC[Nxy2[8].X,Nxy2[8].Y].value xor BitD;
      MasC[XYCur2.X,XYCur2.Y].value:= Sled2;
```

If the unit method of the highest cell of the neighborhood is chosen, then at the even step the coordinates of the current cell are set to zero, and at the even step the unit.

```
if RadioButton2.Checked then
      begin
         Fl:= False;
         for I:= 8 downto 1 do
         begin
            if q mod 2 = 0 then
              begin
                if MasC[Nxy[i].X,Nxy[i].Y].value = 0 then
                  begin
                    XYCur.X:= Nxy[i].X;
                    XYCur.Y:= Nxy[i].Y;
                    Fl:= True;
                  end;
                If Fl Then Break;
              end
            else
              begin
                if MasC[Nxy[i].X,Nxy[i].Y].value = 1 then
                  begin
                    XYCur.X:= Nxy[i].X;
                    XYCur.Y:= Nxy[i].Y;
                    Fl:= True;
                  end;
                If Fl Then Break;
              end;
         end;
         Fl:= False;
```

```
        for I:= 8 downto 1 do
        begin
          if q mod 2 = 0 then
            begin
              if MasC[Nxy2[i].X,Nxy2[i].Y].value = 1 then
                begin
                  XYCur2.X:= Nxy2[i].X;
                  XYCur2.Y:= Nxy2[i].Y;
                  Fl:= True;
                end;
              If Fl Then Break;
            end
          else
            begin
              if MasC[Nxy2[i].X, Nxy2[i].Y].value = 0 then
                begin
                  XYCur2.X:= Nxy2[i].X;
                  XYCur2.Y:= Nxy2[i].Y;
                  Fl:= True;
                end;
              If Fl Then Break;
            end;
        end;
      end;
```

The new coordinates of the cells in the neighborhood and write the new cell statuses into arrays are determined.

```
If XYCur.X-1 < 0 Then Nxy[1].X:= n - 1 else Nxy[1].X:= XYCur.X
- 1;
     If XYCur.Y-1<0 Then Nxy[1].Y:=m - 1 else Nxy[1].Y:=
XYCur.Y - 1;
     Nxy[2].X:=XYCur.X;
     If XYCur.Y - 1 < 0 Then Nxy[2].Y:= m - 1 else Nxy[2].Y:=
XYCur.Y - 1;
     If XYCur.X + 1 > n - 1 Then Nxy[3].X:= 0 else Nxy[3].X:=
XYCur.X+1;
     If XYCur.Y - 1 < 0 Then Nxy[3].Y:= m - 1 else Nxy[3].Y:=
XYCur.Y - 1;
     If XYCur.X + 1 > n - 1 Then Nxy[4].X:= 0 else Nxy[4].X:=
XYCur.X +1;
     Nxy[4].Y:= XYCur.Y;
     If XYCur.X + 1 > n - 1 Then Nxy[5].X:= 0 else Nxy[5].X:=
XYCur.X + 1;
     If XYCur.Y + 1 > m - 1 Then Nxy[5].Y:= 0 else Nxy[5].Y:=
XYCur.Y + 1;
     Nxy[6].X:= XYCur.X;
     If XYCur.Y + 1 > m - 1 Then Nxy[6].Y:= 0 else Nxy[6].Y:=
```

```
XYCur.Y + 1;
     If XYCur.X - 1 < 0 Then Nxy[7].X:= n - 1 else Nxy[7].X:=
XYCur.X - 1;
     If XYCur.Y + 1 > m - 1 Then Nxy[7].Y:= 0 else Nxy[7].Y:=
XYCur.Y + 1;
     If XYCur.X - 1 < 0 Then Nxy[8].X:= n - 1 else Nxy[8].X:=
XYCur.X - 1;
     Nxy[8].Y:= XYCur.Y;
     If XYCur2.X - 1 < 0 Then Nxy2[1].X:=n - 1 else
Nxy2[1].X:= XYCur2.X - 1;
     If XYCur2.Y - 1 < 0 Then Nxy2[1].Y:= m - 1 else
Nxy2[1].Y:= XYCur2.Y - 1;
     Nxy2[2].X:= XYCur2.X;
     If XYCur2.Y - 1 < 0 Then Nxy2[2].Y:= m - 1 else
Nxy2[2].Y:= XYCur2.Y - 1;
     If XYCur2.X + 1 > n - 1 Then Nxy2[3].X:= 0 else
Nxy2[3].X:= XYCur2.X + 1;
     If XYCur2.Y - 1 < 0 Then Nxy2[3].Y:= m - 1 else
Nxy2[3].Y:= XYCur2.Y - 1;
     If XYCur2.X + 1 > n - 1 Then Nxy2[4].X:= 0 else
Nxy2[4].X:= XYCur2.X + 1;
     Nxy2[4].Y:= XYCur2.Y;
     If XYCur2.X + 1 > n - 1 Then Nxy2[5].X:= 0 else
Nxy2[5].X:= XYCur2.X + 1;
     If XYCur2.Y + 1 > m - 1 Then Nxy2[5].Y:= 0 else
Nxy2[5].Y:= XYCur2.Y + 1;
     Nxy2[6].X:= XYCur2.X;
     If XYCur2.Y + 1 > m - 1 Then Nxy2[6].Y:= 0 else
Nxy2[6].Y:= XYCur2.Y + 1;
     If XYCur2.X - 1 < 0 Then Nxy2[7].X:= n - 1 else
Nxy2[7].X:= XYCur2.X - 1;
     If XYCur2.Y + 1 > m - 1 Then Nxy2[7].Y:= 0 else
Nxy2[7].Y:= XYCur2.Y + 1;
     If XYCur2.X - 1 < 0 Then Nxy2[8].X:= n - 1 else
Nxy2[8].X:= XYCur2.X - 1;
     Nxy2[8].Y:= XYCur2.Y;
     MasC[XYCur.X,XYCur.Y].status:= 255;
     MasC[Nxy[4].X,Nxy[4].Y].status:= 4;
     MasC[Nxy[2].X,Nxy[2].Y].status:= 2;
     MasC[Nxy[1].X, Nxy[1].Y].status:= 1;
     MasC[Nxy[3].X, Nxy[3].Y].status:= 3;
     MasC[Nxy[5].X, Nxy[5].Y].status:= 5;
     MasC[Nxy[6].X, Nxy[6].Y].status:= 6;
     MasC[Nxy[7].X, Nxy[7].Y].status:= 7;
     MasC[Nxy[8].X, Nxy[8].Y].status:= 8;
     MasC[XYCur2.X, XYCur2.Y].status:= 255;
     MasC[Nxy2[4].X, Nxy2[4].Y].status:= 4;
     MasC[Nxy2[2].X, Nxy2[2].Y].status:= 2;
     MasC[Nxy2[1].X, Nxy2[1].Y].status:= 1;
```

```
MasC[Nxy2[3].X, Nxy2[3].Y].status:= 3;
MasC[Nxy2[5].X, Nxy2[5].Y].status:= 5;
MasC[Nxy2[6].X, Nxy2[6].Y].status:= 6;
MasC[Nxy2[7].X, Nxy2[7].Y].status:= 7;
MasC[Nxy2[8].X, Nxy2[8].Y].status:= 8
```

Further, if the "accelerate" flag is off, then visualization of the results of the CA's work is organized. Listing below.

```
Cells[XYCur.X,XYCur.Y]:= IntToStr(Sled);
Cells[XYCur2.X,XYCur2.Y]:= IntToStr(Sled2);
Canvas.Brush.Color:= clBlue;
Canvas.FillRect(CellRect(XYCur.X,XYCur.Y));
Canvas.Brush.Color:= clGray;
Canvas.FillRect(CellRect(XYCur2.X,XYCur2.Y));
Canvas.Font.Color:= clWhite; Canvas.
TextOut(CellRect(XYCur.X, XYCur.Y).Left + 5, CellRect(XYCur.X,
XYCur.Y).Top + 2, Cells[XYCur.X, XYCur.Y]);
    Canvas.TextOut(CellRect(XYCur2.X, XYCur2.Y).Left +
5,CellRect(XYCur2.X, XYCur2.Y).Top + 2, Cells[XYCur2.X,
XYCur2.Y]);
Canvas.Brush.Color:= ColCel;
Canvas.FillRect(CellRect(Nxy[4].X, Nxy[4].Y));
Canvas.FillRect(CellRect(Nxy[2].X, Nxy[2].Y));
Canvas.FillRect(CellRect(Nxy[1].X, Nxy[1].Y));
Canvas.FillRect(CellRect(Nxy[3].X, Nxy[3].Y));
Canvas.FillRect(CellRect(Nxy[5].X, Nxy[5].Y));
Canvas.FillRect(CellRect(Nxy[6].X, Nxy[6].Y));
Canvas.FillRect(CellRect(Nxy[7].X, Nxy[7].Y));
Canvas.FillRect(CellRect(Nxy[8].X, Nxy[8].Y));
Canvas.FillRect(CellRect(Nxy2[4].X, Nxy2[4].Y));
Canvas.FillRect(CellRect(Nxy2[2].X, Nxy2[2].Y));
Canvas.FillRect(CellRect(Nxy2[1].X, Nxy2[1].Y));
Canvas.FillRect(CellRect(Nxy2[3].X, Nxy2[3].Y));
Canvas.FillRect(CellRect(Nxy2[5].X, Nxy2[5].Y));
Canvas.FillRect(CellRect(Nxy2[6].X, Nxy2[6].Y));
Canvas.FillRect(CellRect(Nxy2[7].X, Nxy2[7].Y));
Canvas.FillRect(CellRect(Nxy2[8].X, Nxy2[8].Y));
Canvas.Font.Color:= clGreen;
Canvas.TextOut(CellRect(Nxy[4].X, Nxy[4].Y).Left +
2, CellRect(Nxy[4].X, Nxy[4].Y).Top + 2, Cells[Nxy[4].X,
Nxy[4].Y]);
    Canvas.TextOut(CellRect(Nxy[2].X, Nxy[2].Y).Left +
2, CellRect(Nxy[2].X, Nxy[2].Y).Top + 2, Cells[Nxy[2].X,
Nxy[2].Y]);
    Canvas.TextOut(CellRect(Nxy[1].X, Nxy[1].Y).Left +
2, CellRect(Nxy[1].X,Nxy[1].Y).Top + 2, Cells[Nxy[1].X,
Nxy[1].Y]);
```

```
      Canvas.TextOut(CellRect(Nxy[3].X, Nxy[3].Y).Left +
2, CellRect(Nxy[3].X, Nxy[3].Y).Top + 2, Cells[Nxy[3].X,
Nxy[3].Y]);
      Canvas.TextOut(CellRect(Nxy[5].X, Nxy[5].Y).Left +
2, CellRect(Nxy[5].X, Nxy[5].Y).Top + 2,Cells[Nxy[5].X,
Nxy[5].Y]);
      Canvas.TextOut(CellRect(Nxy[6].X, Nxy[6].Y).Left +
2, CellRect(Nxy[6].X, Nxy[6].Y).Top + 2, Cells[Nxy[6].X,
Nxy[6].Y]);
      Canvas.TextOut(CellRect(Nxy[7].X, Nxy[7].Y).Left +
2, CellRect(Nxy[7].X, Nxy[7].Y).Top + 2, Cells[Nxy[7].X,
Nxy[7].Y]);
      Canvas.TextOut(CellRect(Nxy[8].X, Nxy[8].Y).Left +
2, CellRect(Nxy[8].X, Nxy[8].Y).Top + 2, Cells[Nxy[8].X,
Nxy[8].Y]);
      Canvas.Font.Color:= clRed;
      Canvas.TextOut(CellRect(Nxy2[4].X, Nxy2[4].Y).Left +
2, CellRect(Nxy2[4].X, Nxy2[4].Y).Top + 2, Cells[Nxy2[4].X,
Nxy2[4].Y]);
      Canvas.TextOut(CellRect(Nxy2[2].X, Nxy2[2].Y).Left + 2,
CellRect(Nxy2[2].X, Nxy2[2].Y).Top + 2, Cells[Nxy2[2].X,Nxy2[2
].Y]);
      Canvas.TextOut(CellRect(Nxy2[1].X, Nxy2[1].Y).Left +
2, CellRect(Nxy2[1].X, Nxy2[1].Y).Top + 2, Cells[Nxy2[1].X,
Nxy2[1].Y]);
      Canvas.TextOut(CellRect(Nxy2[3].X, Nxy2[3].Y).Left +
2, CellRect(Nxy2[3].X, Nxy2[3].Y).Top + 2, Cells[Nxy2[3].X,
Nxy2[3].Y]);
      Canvas.TextOut(CellRect(Nxy2[5].X, Nxy2[5].Y).Left +
2, CellRect(Nxy2[5].X, Nxy2[5].Y).Top + 2, Cells[Nxy2[5].X,
Nxy2[5].Y]);
      Canvas.TextOut(CellRect(Nxy2[6].X, Nxy2[6].Y).Left +
2, CellRect(Nxy2[6].X, Nxy2[6].Y).Top + 2, Cells[Nxy2[6].X,
Nxy2[6].Y]);
      Canvas.TextOut(CellRect(Nxy2[7].X, Nxy2[7].Y).Left +
2, CellRect(Nxy2[7].X, Nxy2[7].Y).Top + 2, Cells[Nxy2[7].X,
Nxy2[7].Y]);
      Canvas.TextOut(CellRect(Nxy2[8].X, Nxy2[8].Y).Left +
2, CellRect(Nxy2[8].X, Nxy2[8].Y).Top + 2, Cells[Nxy2[8].X,
Nxy2[8].Y]);
```

CONCLUSION

The results of this chapter complement the results of the previous chapter. New paradigms for AKA with several active cells that have different laws of functioning are described. Studies have allowed us to create ACA in which active cells do not enter the cycles and move along the ACA field for a long time. The results obtained in this chapter allowed us to proceed to the next step, in which ACA with many active cells with different active states are formed.

REFERENCES

Belan, S. N., & Motornyuk, R. L. (2013). Extraction of characteristic features of images with the help of the radon transform and its hardware implementation in terms of cellular automata. *Cybernetics and Systems Analysis*, *49*(1), 7–14. doi:10.100710559-013-9479-2

Bilan, S. (2017). *Formation Methods, Models, and Hardware Implementation of Pseudorandom Number Generators: Emerging Research and Opportunities*. IGI Global.

Bilan, S., Bilan, M., & Bilan, S. (2015). Novel pseudorandom sequence of numbers generator based cellular automata. *Information Technology and Security*, *3*(1), 38–50.

Bilan, S., Bilan, M., Motornyuk, R., Bilan, A., & Bilan, S. (2016). Research and Analysis of the Pseudorandom Number Generators Implemented on Cellular Automata. *WSEAS Transactions on Systems*, *15*, 275–281.

Bilan, S., Motornyuk, R., & Bilan, S. (2014). Method of Hardware Selection of Characteristic Features Based on Radon Transformation and not Sensitive to Rotation, Shifting and Scale of the Input Images. *Advances in Image and Video Processing*, *2*(4), 12–23. doi:10.14738/aivp.24.392

Motornyuk, R. L. (2013). *Computer-aided methods for identifying images of moving objects based on cellular automata with a hexagonal coating*. Dissertation for the degree of candidate of technical sciences (UDC 004.932: 519.713 (043.3)). Kiev: SUIT.

Chapter 5
Models and Paradigms of Cellular Automata With a Variable Set of Active Cells

ABSTRACT

The chapter describes the functioning model of an asynchronous cellular automaton with a variable number of active cells. The rules for the formation of active cells with new active states are considered. Codes of active states for the von Neumann neighborhood are presented, and a technique for coding active states for other forms of neighborhoods is described. Several modes of operation of asynchronous cellular automata from the point of view of the influence of active cells are considered. The mode of coincidence of active cells and the mode of influence of neighboring active cells are considered, and the mode of influence of active cells of the surroundings is briefly considered. Algorithms of cell operation for all modes of the cellular automata are presented. Functional structures of cells and their CAD models are constructed.

INTRODUCTION

In the modern theory of CA, many dynamic processes are described and many of their models have already been built. Much attention is paid to the processes of the appearance of new objects and the disappearance of existing objects. Objects can be represented by a single cell or their combination. As

DOI: 10.4018/978-1-7998-2649-1.ch005

a rule, such processes for each CA cell are determined by the state of the neighborhood cells, which can vary according to the given transition rules. The processes of interaction of several separate selected cells in the CA field are practically not described. For this interaction, the theoretical positions for ACA with active cells, described in previous chapters, were selected. Moreover, new paradigms allow the formation of new cells with new properties based on the existing properties of old cells that interact.

The chapter describes the functioning model of an asynchronous cellular automaton with a variable number of active cells. The rules for the formation of active cells with new active states are considered. Codes of active states for the von Neumann neighborhood are presented, and a technique for coding active states for other forms of neighborhoods is described. Several modes of operation of asynchronous cellular automata from the point of view of the influence of active cells are considered. The mode of coincidence of active cells and the mode of influence of neighboring active cells are considered, and the mode of influence of active cells of the surroundings is briefly considered. Algorithms of cell operation for all modes of the cellular automata are presented. Functional structures of cells and their CAD models are constructed.

MODEL AND ALGORITHMS FOR THE FUNCTIONING OF CELLULAR AUTOMATA WITH SEVERAL ACTIVE CELLS IN THE INFLUENCE MODE OF THE NEIGHBORHOOD CELLS

In the previous sections, ACA with one and several active cells was considered. In these ACAs, the number of active cells does not change. Also, they do not change LTF and LSF. It was shown that only active cells change their state at each time step or perform another LSF, which is different from the LSF that other inactive cells perform.

This chapter discusses the models and structures of ACA, which tend to change the number of active cells. In this case, ACA can use various algorithms for the formation and removal of active cells. This situation is partially considered in the work (Bilan, Bilan, & Bilan, 2017). Several examples of changes in the number of active cells are described.

In general, changes in active ACA cells can occur in two modes:

1. The mode of influence of the cells of the neighborhood;

2.   The mode of influence of active cells.

The standby mode is characterized by the fact that the cell is not active and is in a logical state of "1" or "0". In the active state, the cell performs LSF and can change its information state according to LSF. The result of LSF execution depends on the values of the arguments that are formed at the outputs of the cells of its neighborhood. The argument can also be the value at the output of the most active cell in the previous time step. The second mode refers to the active cell. The second mode is similar to the active cell mode for an ACA with one active cell, which was discussed in the previous section.

The mode of the second active cell is similar to the mode of the first active cell. However, the difference is that the second active cell performs another LSF.

The fourth mode corresponds to a cell that is in the states of the first and second active cells at the current time step. In this mode, an active cell cannot simultaneously execute two different LSFs, so each cell has one informational state (one informational output). In such a situation, the following submodes may occur:

- LSF with high priority;
- Each of the LSFs is performed at a specific time step;
- An LSF is executed, the argument of which is the resulting signals of the first and second LSF.

In the influence mode of the neighborhood cells, active cells appear and disappear spontaneously (pseudo-randomly), i.e. the number of active cells may increase sharply, or may decrease sharply. In this case, the fact that only active cells change their basic information state is taken into account.

The following mathematical model is used to describe the evolution of the behavior of such an ACA

$$S(t+1) = \begin{cases} f\left[x_1(t),\ldots,x_n(t)\right], if \ f_{act}\left[x_1(t),\ldots,x_n(t)\right] = 1 \\ 0 \ in \ other \ case \end{cases} \tag{15}$$

where

$f\left[x_1(t),\ldots,x_n(t)\right]$ – LSF of cells;

$f_{act}\left[x_1(t), \ldots, x_n(t)\right]$ – local cell function, which determines the state of cell activity in the next time step.

To implement the model (15), several modes can be used: the mode without memory and the mode with memory.

In memoryless mode, the cell is in the active state only at time steps, when $f_{act}\left[x_1(t), \ldots, x_n(t)\right] = 1$. If this function for one cell takes the value of logical "1" at several steps in a row, then at these steps this cell is in an active state. An example of the functioning of such a cellular automata in Figure 1 is shown.

Figure 1. An example of the functioning of ACA with a variable number of active cells

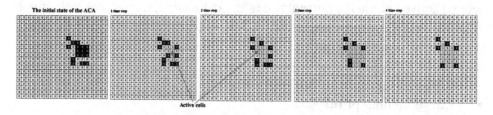

In this example, an ACA is presented in which the neighborhood of each cell is realized as the Moore neighborhood, and the cell becomes active if the following majority function

$$S_{act}(t+1) = \begin{cases} 1, if \sum_{i=1}^{n} x_i(t) > \dfrac{n}{2}, \\ 0\, in\, other\, case \end{cases} \quad (16)$$

where n=8 – the number of cells in the neighborhood of Moore.

At the initial time, the ACA contains an equal number of cells having logical states of "1" and "0". The LSF uses the OR function. As can be seen from Figure 1 the number of active cells increases at subsequent time steps in the evolution of ACA.

If you use the AND function as LSF, then the number of active cells decreases at each time step of the ACA evolution (Figure 2).

Figure 2. An example of the functioning of an ACA with a variable number of active cells and the LSF used as an AND function

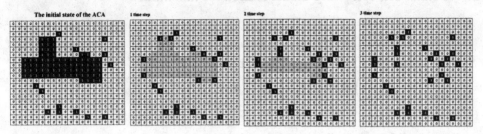

It can be seen from these examples that there is no need to generate special signals for setting the cell in an active state, as well as for the transition of a cell from an active state to a resting state.

In memory mode, the cell goes into active state at the current time step if $f_{act}\left[x_1(t),...,x_n(t)\right]=1$. However, in the next time step, the active cell remains active even if $f_{act}\left[x_1(t),...,x_n(t)\right]=0$. In this mode, the active state of the cell is maintained at all subsequent time steps in the evolution of ACA. There is a fact that for most LSFs, the number of active cells increases at each time step in the evolution of ACA. This situation often leads to the full activity of the entire ACA. In this case, the ACA goes into SCA since all the cells of the CA perform the same function.

If there is a need to use ACA with a variable number of active cells, while the number of active cells can not only increase, but also decrease, then the special signal generation mode is used to zero the active state of the cells. For this, a special activation function is used, which puts the cell in an inactive state

$$S_{act}(t+1) = \begin{cases} 1, if\ f_{act}\left[x_1(t),...,x_n(t)\right]=1 \\ 0, if\ f_0\left[x_1(t),...,x_n(t)\right]=1 \\ S_{act}(t), 0\ in\ other\ case \end{cases} \quad (17)$$

where $f_0\left[x_1(t),...,x_n(t)\right]$ – active state reset function.

For this model, it is important to select the priority of the activating function. If you use one activating function, then for the activation and for zeroing can be used different values of one function. Using multiple functions requires prioritization. As a rule, the regime is used when the cell was in an active state. The following model is used for this mode.

$$S_{act}(t+1) = \begin{cases} 1, if\ f_{act}\big[x_1(t),...,x_n(t)\big] = 1 \\ 0, if\ f_0\big[x_1(t),...,x_n(t)\big] = 1\ and\ S_{act}(t) = 1, \\ S_{act}(t), 0\ in\ other\ case \end{cases} \qquad (18)$$

It is possible to introduce additional conditions and expand experimental studies of the behavior of ACAs, as well as study the behavior of such ACAs for different LSFs. An example of the behavior of ACA in the active memory mode on Figure 3 is shown.

Figure 3. An example of the functioning of the ACA in the mode of memory of the active cells

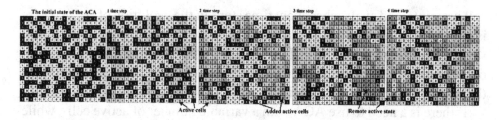

For this example, LSF is used, which is implemented by the XOR function. The majority function is used as the activating function $f_{act}\big[x_1(t),...,x_n(t)\big]$, and a function of the form is used as the zeroing function $f_0\big[x_1(t),...,x_n(t)\big] < 3$.

As can be seen from Figure 3, the number of active cells (highlighted in yellow) at each time step either increases (added active cells at each time step are highlighted in orange) or decreases (cells with deleted active states are highlighted in blue). In this mode, active cells appear or disappear at each time step, depending on the informational states of all cells. It is possible that in one time step all cells become active or inactive. If in the ACA all cells become inactive and no active cells appear at the next time step, then the ACA becomes inactive. It conditions do not change.

MODELS AND ALGORITHMS FOR THE FUNCTIONING OF ACA WITH A VARIABLE SET OF ACTIVE CELLS IN THE MODE OF INFLUENCE OF ACTIVE CELLS

In the previous section, ACAs with a variable number of active cells were considered, in which the state of all cells, the shape of the neighborhood, LSF, and the activation function have the main influence on evolution. In this case, activation cells do not contribute to the appearance of new active cells.

There are options for the implementation of ACA with a variable number of active cells, in which new active cells appear and disappear as a result of the interaction of several active cells. Compared to the ACA described in the previous section, this ACA requires the initial installation of active cells.

There are a different number of options for the interaction of active cells, as a result of which active cells appear (are born) or disappear (die). As previously described, active cells transmit an active signal to one of the cells of their own neighborhood at the next time step of the ACA. At the next time step, one of the cells in the neighborhood of the active cell becomes active, and the active cell at the previous time step is zeroed.

Thus, we will talk about the transmission of the active signal, as about the movement of the active cell in the ACA field (Bilan, 2017; Bilan, et al 2016). In fact, all cells remain in place, and the state of the cell that characterizes the properties of activity (for example, color, state at the exit, etc.) moves. However, the work will talk about the movement of active cells in the ACA field.

As in all previous ACA models described, the active cell fulfills the given LSF. The formula (5.1) is applied to the cells of such ACAs. Active cells move in the ACA field and can perform the same or different LSFs. In addition, different LTFs are used for each individual active state.

Active cells can interact with each other (influence each other) in the following cases:

1. Active cells coincided in one cell at the current time step.
2. Active cells are neighbors, i.e. are one of the neighborhood cells.
3. Active cells have common cells in the neighborhood.

There are also modes in which only one of the active cells can influence the other (domination). An example of this dominance is prioritization of the LSF results of the corresponding active cells. If the quantitative LSF of one active cell exceeds the quantitative LSF of another active cell, then the cell with the larger value dominates the cell with the smaller value.

In the previous chapter, ACA with several active cells was considered. Active cells with different active states moved along the ACA field. However, they have no effect on each other. In doing so, they perform LSFs, which may be different for each active cell. LSF options for each active cell are described in Chapter 4. Each active cell can perform the same LSF, and can also perform different LSFs. However, the question arises. What LSF will the new active cell perform?

The answer to this question is the following options:

1. The new active cell will perform the same LSF as the two interacting active cells;
2. The new active cell will execute the predefined LSF;
3. The new active cell will perform LSF, which depends on a combination of interacting active cells.

The first option is acceptable for ACA, in which all active cells perform the same LSF. Such an AKA is the easiest to implement. The operation of the cell of such an AKA is described by the following model

$$S(t+1) = \begin{cases} f\big[x_1(t),\dots,x_n(t)\big], if\ X_{i,act(j)} = 1 \\ S(t)\ in\ other\ case \end{cases},$$

(19)

where $X_{i,act(j)}$ – signal on the i-th active output from the i-th output of the j-th active cell.

An example of the implementation of LSF by active cells in this mode in Figure 4 is presented.

Figure 4. An example of the evolution of ACA in the execution mode of one LSF by all active cells

For the formation of a new cell and visualization, separation and combination of active cells by color was used. In the presented example, the green and yellow active cells form a new blue cell that performs the same LSF. Near to each active cell are three cells of the Moore neighborhood that encode the active state and direction of movement of the active signal. In the third step, the two active cells coincided and formed a new active cell with a new active state. The transition of a cell to an active state in this mode is described by the following model for two active cells

$$S_{act}(t+1) = \begin{cases} S_{act(1)}, if\ f_{act(1)}\left[x_1(t),...,x_n(t)\right] = 1, f_{act(2)}\left[x_1(t),...,x_n(t)\right] = 0 \\ S_{act(2)}, if\ f_{act(1)}\left[x_1(t),...,x_n(t)\right] = 0, f_{act(2)}\left[x_1(t),...,x_n(t)\right] = 1 \\ S_{act(3)}, if\ f_{act(1)}\left[x_1(t),...,x_n(t)\right] = 1, f_{act(2)}\left[x_1(t),...,x_n(t)\right] = 1 \\ 0, in\ the\ other\ case \end{cases}$$

(20)

The model is valid for the initial two active cells and does not take into account the fact that the number of active cells increases. Situations may arise when the formed third cell meets the first or second active cells at a specific time step. Then again, the same question arises. What LSF and which LTF will the new formed active cell perform? For the first mode, the LSF is the same, and the LTF should be formed in accordance with the specified activation function.

One of the simplest and most accessible modes of operation of an ACA with a variable number of active cells is the use of state signals of cells in the neighborhood of each interacting active cell. The number of neighborhood cells used is limited by the shape and structural neighborhood. For example, for a von Neumann neighborhood, two neighborhood cells are used, with which you can specify four directions of active signal transmission. In this case, the number of possible active cells that may be present in the ACA is $A_4^2 = 12$.

In general, the number of cells that encode an active state is determined from the formula

$$k_{act} = \sqrt[a]{n},$$

(21)

where

k_{act} – number of cells encoding an active state;

n – the number of cells in the neighborhood of the active cell;

a – the number of possible cells that give n binary truth table sets.

In general, the number of possible active states is determined by the following formula

$$A_{st} = A_n^k = \frac{n!}{(n-k)!}.$$ (22)

In this approach, different active cells can transmit the active signal in the same direction at a certain time step in the evolution of ACA. However, when several active cells interact with each other, a new active state is formed under the indication of the resulting function. Code resulting from the interaction of two activating functions. One of the options for determining a new active state is to compile a table of possible results of the active function of the cell in which the new active state is formed. For the neighborhood of von Neumann, a table is compiled, which is represented by table 1.

Table 1 shows that a new active state is formed from two disparate neighborhood cells for two interacting active cells. The value of a_i indicates the location of the cell in the neighborhood (Figure 5).

This is the most convenient and understandable option for coding new active states. Moreover, the function can be described in terms of set theory

$$C_{new} = (A \cup B) \setminus (A \cap B) = (A \setminus B) \cup (A \setminus B).$$ (23)

According to model (20), the new active state $S_{act(3)}$ arises as a result of the interaction of two cells with different active states. This condition is described by the following model

$$S_{act(3)} = \begin{cases} 1, if \ A_1 = A_2 = m \ and \ (A \cup B) \setminus (A \cap B) = C, C = m \\ 0 \ in \ other \ case \end{cases},$$ (24)

where

Table 1. Table for the formation of new active states for the von Neumann neighborhood

| Codes of Active States of Interacting Cells | | | | New Active State | |
| First Active Cell | | Second Active Cell | | | |
Active State Number	Code of Active State	Active State Number	Code of Active State	Active State Number	Code of Active State
1	a_1a_2	1	a_1a_2		0
	a_1a_2	2	a_1a_3	4	a_2a_3
	a_1a_2	3	a_1a_4	5	a_2a_4
	a_1a_2	4	a_2a_3	2	a_1a_3
	a_1a_2	5	a_2a_4	3	a_1a_4
	a_1a_2	6	a_3a_4		-
2	a_1a_3	1	a_1a_2	4	a_2a_3
	a_1a_3	2	a_1a_3		0
	a_1a_3	3	a_1a_4	6	a_3a_4
	a_1a_3	4	a_2a_3	1	a_1a_2
	a_1a_3	5	a_2a_4		-
	a_1a_3	6	a_3a_4	3	a_1a_4
3	a_1a_4	1	a_1a_2	5	a_2a_4
	a_1a_4	2	a_1a_3	6	a_3a_4
	a_1a_4	3	a_1a_4		0
	a_1a_4	4	a_2a_3		-
	a_1a_4	5	a_2a_4	1	a_1a_2
	a_1a_4	6	a_3a_4	2	a_1a_3
4	a_2a_3	1	a_1a_2	2	a_1a_3
	a_2a_3	2	a_1a_3	1	a_1a_2
	a_2a_3	3	a_1a_4		-
	a_2a_3	4	a_2a_3		0
	a_2a_3	5	a_2a_4	6	a_3a_4
	a_2a_3	6	a_3a_4	5	a_2a_4
5	a_2a_4	1	a_1a_2	3	a_1a_4
	a_2a_4	2	a_1a_3		-
	a_2a_4	3	a_1a_4	1	a_1a_2
	a_2a_4	4	a_2a_3	6	a_3a_4
	a_2a_4	5	a_2a_4		0
	a_2a_4	6	a_3a_4	4	a_2a_3
6	a_3a_4	1	a_1a_2		-
	a_3a_4	2	a_1a_3	3	a_1a_4
	a_3a_4	3	a_1a_4	2	a_1a_3
	a_3a_4	4	a_2a_3	5	a_2a_4
	a_3a_4	5	a_2a_4	4	a_2a_3
	a_3a_4	6	a_3a_4		0

Figure 5. Cell coding for von Neumann and Moore neighborhoods

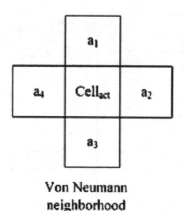

Von Neumann
neighborhood

Neighborhood of Moore

A_1 and A_2 – many neighborhood cells that encode the first and second active
state;

M – the number of cells in the neighborhood of the active cell that encodes
the active state.

The used function (24) is easily implemented both programmatically and
by hardware. Moreover, from table 1 it is seen that there are combinations
when the result at the output is no active state. In this case, we can talk about
the zeroing of active states or that there are no changes. If two identical active
states occur in the same cell and interact, then these states are reset, since the
result of the fulfillment of formula (23) is an empty set $\emptyset$. If there are two
meeting active states in which fulfillment of formula (23) gives a set with
a higher power, then no changes occur. An example of the evolution of the
ACA in this mode is presented in Figure 6.

*Figure 6. An example of the evolution of ACA in the mode of influence of active
cells with a neighborhood of von Neumann and one LSF*

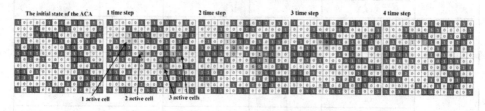

This figure shows the appearance and disappearance of active cells. Each active state of the cell is represented by a specific color (yellow is the first active state, green is the second active state, blue is the third active state). Near each active cell, neighborhood cells are selected that encode the corresponding active state. In the coincidence mode of active cells in one cell, the probability of coincidence of several states in one cell is low. In this case, only one of the possible cell matching options is considered.

Situations are also possible when three or more active states coincide in one cell. For example, if three different active states coincide, then three more new active cells are born in the next time step. If the new active states are different, then they continue to function at subsequent time steps (Figure 7).

Figure 7. An example of the functioning of the ACA in the mode of influence of active cells in the case of the coincidence of three different active states. Von Neumann neighborhood and one LSF are used.

The example presented in Figure 7 describes one match option in the third time step. In the first, second and third time steps, cells are allocated near each active cell, encode the active state of the corresponding cell. At the fourth time step, two active cells with two new active states are visible (a_1a_3 and a_2a_4). Also, the pink cell has three active states (a_1a_2; a_2a_3; a_3a_4). In the fifth time step, two active cells with the same new active states as in the fourth time step should also appear. Moreover, their location depends on the state of the codes that encode these active states. However, at the next time step, several active cells with the same active states may form. These cells with the same active states do not live long (one time step) and disappear at the next time step (Figure 8).

In this example, four active cells meet in one cell in the second time step, where active cells with new active states are formed (two new active cells have active states a_1a_2 and two new active cells have active states a_3a_4). These cells nullify each other and active cells with old active states remain a_2a_3; a_1a_3; a_2a_4; a_1a_4. At the third time step, these states combine two in one cell (a_2a_3;

Figure 8. An example of the functioning of AKA with the disappearance of new active cells for the neighborhood of von Neumann and one LSF

a_2a_4 and a_1a_3; a_1a_4). In the fourth time step, these cells form two new active cells with identical active states a_1a_4. Since the new active cells are located in different AKA cells, they do not nullify.

The same examples can be applied to ACAs, which are organized for different forms of neighborhoods and for different LSFs. The choice of neighborhood forms and LSF depends on the task and process that are being modeled.

The algorithm of the ACA cell in the matching mode of active cells consists of the following steps.

1. At the initial moment of time, the cell is set to the main information state (logical "0" or "1").
2. Setting cell active states.
3. Analysis of the cell's own state.
    3.1. The cell does not have an active state. Go to step 4.
    3.2. The cell is in an active state. Go to step 7.
4. Signal analysis at active cell inputs.
    4.1. At the active inputs of the cell there are logical "0". Go to step 6.
    4.2. At one or more active inputs, logic 1 signals are present. Go to step 5.
5. Transition of the cell to the corresponding active states, indicated by the codes of neighboring active cells.
6. Shutting down a cell at a selected time step.
7. Analysis of the own active states.
    7.1. The cell is in one active state. Go to step 8.
    7.2. The cell is in several active states. Go to step 9.
8. The current active state of the cell is reset. Go to step 4.
9. The formation of new active states in accordance with a given encoding and active function.

10. Zeroing the current active states and forming active signals at the corresponding active outputs. Go to step 6.

MODELS AND ALGORITHMS FOR THE FUNCTIONING OF ACA WITH A VARIABLE SET OF ACTIVE CELLS IN THE INFLUENCE MODE OF NEIGHBORING ACTIVE CELLS

This section describes the appearance of new active cells under the influence of neighboring active cells. In this mode, new active cells at the next time step appear when at the current time step two active cells are adjacent and belong to the cells in the neighborhood of these active cells. Each of the active cells is a cell in the neighborhood of another active cell. For example, for a von Neumann neighborhood, four active cells can be adjacent at the same time, and eight for a Moore neighborhood (Figure 9).

Figure 9. An example of influencing active cells located in the neighborhood of each cell

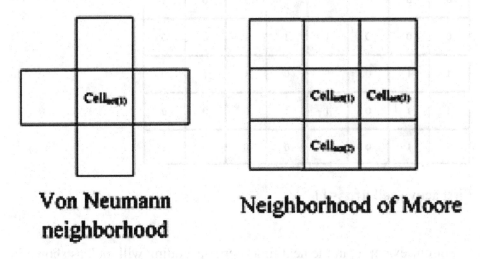

Von Neumann neighborhood

Neighborhood of Moore

The question arises. In which cell will a new active cell form? There can be many answers and options. For example, a new active cell can be formed in an active cell, which is located on the right or top. Let us dwell on the option when the codes of two active neighboring states are compared and the active state code with a large value is selected. For example, the code of the first active state is $a_1^1 a_2^1 = 01$ (for the von Neumann neighborhood), and the code of the second active state is $a_1^2 a_3^2 = 00$, then a new active cell is formed in the cell with the first active state on the next time step. The active state code of the new active cell will correspond $a_2 a_3$. The superscripts indicate that codes of active states are selected from different neighborhoods (Figure 10). These neighborhoods are neighborhoods of interacting active cells.

Figure 10. An example of coding the active states of two interacting neighboring cells for a von Neumann neighborhood

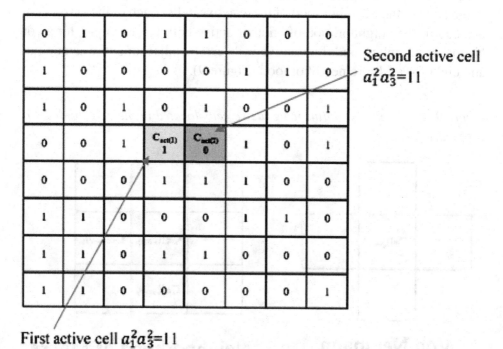

Second active cell
$a_1^2 a_3^2 = 11$

First active cell $a_1^2 a_3^2 = 11$

For our example, in the next time step, the coding will look as shown in Figure 11.

From this example, it is clearly seen that at the next time step, two more new active cells are formed. Their active states will be encoded as $a_1^5 a_3^5$ and $a_1^4 a_2^4$ (Figure 12).

At this time step, the new fifth active cell does not appear because the codes of the first and second active cells coincide and it is impossible to

Figure 11. An example of the formation and coding of a new active cell in the interaction of neighboring active cells

0	1	0	1	1	0	0	0
1	0	0	0	0	1	1	0
1	0	1	$C_{act(1)}$ 0	1	0	0	1
0	0	1	$C_{act(2)}$ 1	$C_{act(3)}$ 0	1	0	1
0	0	0	1	1	1	0	0
1	1	0	0	0	1	1	0
1	1	0	1	1	0	0	0
1	1	0	0	0	0	0	1

$$a_1^1 a_2^1 = 01$$
$$a_1^2 a_3^2 = 01$$
$$a_2^3 a_3^3 = 11$$

determine the location of the fifth active cell. The emplacement of active cells coincides with the second and fourth active states. Since the first and fourth active cells have the same active state $\left(a_1^1 a_2^1 = a_1^4 a_2^4\right)$, then at the next time step they disappear (Figure 13). Active cells remain with the second and third active states $\left(a_1^2 a_3^2 ; a_2^3 a_3^3\right)$. These cells participate in the formation of a new active cell with an active state $a_1^5 a_2^5$ (Figure 13).

In this situation, the balance is maintained until it leaves the active cells from the neighborhood of other active cells and moves away from the created group. The number of active states in such a group will be limited until an active cell that comes from another group with other active states begins to interact with it.

Such interaction of neighboring active cells has a behavior that is characterized by the formation of groups of active cells with similar active states. However, as mentioned earlier, this is one of the options for the interaction of neighboring active cells. The following model can represent the active state of the cell.

Figure 12. The formation of new active cells in the adjacent interaction

0	1	0	1	1	0	0	0
1	0	0	0	0	1	1	0
1	0	1	1	$C_{act(1)}$ 1	0	0	1
0	0	1	$C_{act(3)}$ 1	$C_{act(2)}$ $C_{act(4)}$ 0	1	0	1
0	0	0	1	1	1	0	0
1	1	0	0	0	1	1	0
1	1	0	1	1	0	0	0
1	1	0	0	0	0	0	1

$$S_{act}\left(t+1\right) = \begin{cases} S_{act(j)}, if\ \exists \forall a_i\left(t\right) = S_{act(j-2)}, S_{act(j-1)} = 1, i = \overline{1,n} \\ S_{act(j-1)}, if\ \exists \forall a_i\left(t\right) \neq S_{act(j-2)}, S_{act(j-1)} = 1 \\ S_{act(j-2)}, if\ \exists \forall a_i\left(t\right) \neq S_{act(j-1)}, S_{act(j-2)} = 1 \\ 0, if\ \exists \forall a_i\left(t\right) \neq S_{act}\left(t\right), S_{act}\left(t\right) = 0 \\ S_{act(k)}, if\ \exists \forall a_i\left(t\right) = S_{act(k)}\left(t\right), a_i^k a_l^k = z \end{cases} \tag{25}$$

where z – the number of the current cell in the neighborhood indicated by the code of the neighboring active cell with the kth active state.

The formula (25) is limited, since it does not take into account the zeroing of neighboring active cells. Also does not take into account several possible active cells in the neighborhood of active cells and transition to several current or new active states is not described.

The ACA cell operation algorithm in this mode can be represented by the following description of steps.

Figure 13. An example of the formation of a fifth active cell with an active state $a_1^5 a_2^5$

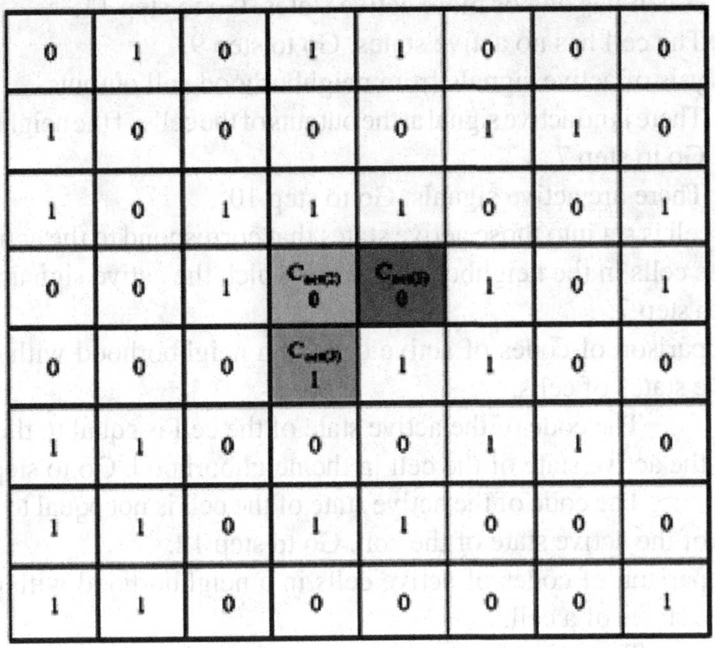

0	1	0	1	1	0	0	0
1	0	0	0	0	1	1	0
1	0	1	1	1	0	0	1
0	0	1	$C_{ac(2)}$ 0	$C_{ac(3)}$ 0	1	0	1
0	0	0	$C_{ac(3)}$ 1	1	1	0	0
1	1	0	0	0	1	1	0
1	1	0	1	1	0	0	0
1	1	0	0	0	0	0	1

1. All ACA cells are set to the main information state.
2. Setting the cell in an active state. At this step, the cell may have the kth active state or several active states. Also, the cell may not have an active state.
3. Neighborhood cell analysis.
   3.1. One or more neighborhood cells are in an active state. Go to step 8.
   3.2. None of the cells in the neighborhood is in an active state. Go to step 4.
4. Analysis of own active state.
   4.1. A cell has one or more active states. Go to step 5.
   4.2. The cell has no active states. Go to step 7.
5. The formation of active signals at the corresponding active outputs of the cell for the transfer of active states to the corresponding cells in the neighborhood. Codes of each active state of the cell indicate the number of cells in the neighborhood (Figure 8).
6. Zeroing previous active cell states.

7.   the end.
8.   Analysis of the cell's own state.
    8.1. A cell has one or more active states. Go to step 11.
    8.2. The cell has no active states. Go to step 9.
9.   Analysis of active signals from neighborhood cell outputs.
    9.1. There is no active signal at the outputs of the cells of the neighborhood. Go to step 7.
    9.2. There are active signals. Go to step 10.
10.   The cell is set into those active states that correspond to the active states of the cells in the neighborhood from which the active signals arrived. Go to step 7.
11.   Comparison of codes of active cells in a neighborhood with codes of active states of cells.
    11.1.   The code of the active state of the cell is equal to the code of the active state of the cell in the neighborhood. Go to step 14.
    11.2.   The code of the active state of the cell is not equal to the code of the active state of the cell. Go to step 12.
12.   Comparison of codes of active cells in a neighborhood with codes of active states of a cell.
    12.1.   The code of the active state of the cell is greater than the code of the active state of the cell in the neighborhood. Go to step 13.
    12.2.   The code of the active state of the cell is less than the code of the active state of the cell in the neighborhood. Go to step 5.
13.   The formation of a new active state of the cell in accordance with the codes of the active states of the cell and the active cells of the neighborhood. Go to step 5.
14.   Zeroing the active state of the cell. Go to step 7.

In the points of the algorithm where we are talking about the analysis of the state of the cells of the neighborhood, a whole procedure is used, which is determined by the built-in algorithm for sorting all the cells. As a result of this built-in algorithm, active cells in the neighborhood and their state codes are determined. To speed up the implementation of the algorithm, initially the presence of active cells in the neighborhood is generally determined. If it is determined that there are active cells in the neighborhood, then enumeration of all cells in the neighborhood begins, if the control cell is active.

The software and hardware implementation of this ACA operating mode is much more complicated than the implementation of the ACA described for the previous mode. Moreover, such ACAs have different evolution under equal initial conditions. In Figure 14 presents the evolution for ACA with the influence of neighboring active cells.

Figure 14. ACA evolution for coincidence and neighborhood modes

In this Figure 14, two active cells with active states that are encoded as a_1a_2 (yellow) a_1a_4 (lilac) are initially used. At the second time step, active cells begin to influence each other and a new active state a_2a_4 (red color of the cell) is formed at the third time step. However, at the third time step, the active states coincide in the same cell a_1a_2 and a_2a_4 (red and yellow coincide). These active cells influence each other and at the third time step two more states are formed. At the fifth time step, the red active states are zeroed and only four active states remain in the four ACA cells. Obviously, the evolution of the ACA for the second mode is different from the ACA operating in the first mode. In the second mode, more new active cells are formed at each time step.

MODELS AND ALGORITHMS OF FUNCTIONING OF AN ACA WITH A VARIABLE SET OF ACTIVE CELLS IN THE INFLUENCE MODE OF CELLS IN THE NEIGHBORHOOD OF ACTIVE CELLS

In the previous sections, ACA functioning algorithms were considered when active cells influence each other by coincidence and if the active cells are adjacent. The third variant of influence may be the variant when the cells belonging to the neighborhood of active cells are influencing (Figure 15).

As can be seen from Figure 15 with the use of the von Neumann neighborhood, only one cell of the neighborhood can be influential for the case when active cells approach one another. For a Moore neighborhood, there can be from one to three influencing cells of the neighborhood of the two nearest active cells. Again, the same question arises. Where will a new

Figure 15. An example of the influence of cells of the neighborhood of von Neumann and Moore

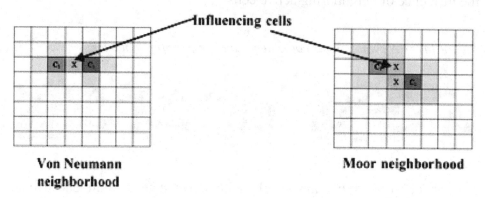

Von Neumann neighborhood

Moor neighborhood

active cell with a new active state be formed (born)? If for the von Neumann neighborhood we can accept that a new active cell will form in the influencing neighborhood, then for the Moore neighborhood it is necessary to establish priorities. It is easy to assume that if one cell is the influencing cell, then a new active cell with a new active state will form in it. If two neighborhood cells are active (these cells are adjacent), then a cell with a higher priority will be selected (for example, the cell that is located at the top of the neighborhood), and if three neighborhood cells are influential, then the middle cell among them will be selected. There may be other cell choices for the formation of new active cells.

The model for the formation of a new active cell for the von Neumann neighborhood has the following form

$$S_{act}(t+1) = \begin{cases} S_{act(3)}, & if \ \sum_{i=1}^{7} Y_{act,i} = 2 \\ S_{act(1)}, & if \ X_{act,(1)}(t) = 1 \\ S_{act(2)}, & if \ X_{act(2)}(t) = 1 \\ 0 \ if \ X_{act,i}(t) = X_{act(2)}(t) = 0 \end{cases}, \tag{26}$$

where $X_{act}(t)$ – the signal at the active input of the considered cell.

If we assume that the active states are encoded according to Table 1 for the von Neumann neighborhood, then an example of the evolution of the ACA for this mode can be presented in the following form (Figure 16).

Figure 16. An example of the evolution of an ACA in the influence mode of neighborhood cells for a von Neumann neighborhood

In the presented example, it can be seen that at the third time step two new identical active cells are formed (the blue color of the cell). However, they are adjacent and disappear in the next time step. In this case, a new active cell appears again (blue color of the cell) at the fourth time step.

Variants are possible when at the next time step all active cells coincide in one cell and then all cells in the neighborhood influence each other simultaneously. In such a situation, four new active cells or one new active cell can form. If one new active cell is to be formed, then it is necessary to establish a priority for each influencing cell of the neighborhood. This priority can be set by executing a specific logical function (for example, XOR function). The result of performing a logical function in the form of a two-bit code indicates the number of the surrounding cell in the neighborhood where a new active cell is formed at the next time step.

Coincidence of all active cells may result from the formation of a new active cell. If in the process of evolution two active cells are adjacent, then in this mode there are no influencing cells and new active cells are not formed.

According to table 1, the active cells can nullify the same active state. Much depends on the state of the neighborhood cells that encode each active state. They set the direction of transmission of the active signal in the next time step. Colonies of the active cells can be created. In addition, separate isolated active cells can form, which, reaching another colony, take part in the formation of new active states.

CONCLUSION

The use of ACA with several active cells allowed to build models for the formation of new cells with new states and transition functions. This approach is implemented on the ACA with active cells that perform different local functions of states and transitions. The possibility of the formation of new active states, which are described by the states of the cells of the neighborhood, is also realized. It is shown that the number of possible different active cells depends on the shape of the neighborhood and the number of cells that form it. The proposed methods and paradigms allow to describe the behavior of living organisms of various forms and levels based on ACA.

REFERENCES

Bilan, S. (2017). *Formation Methods, Models, and Hardware Implementation of Pseudorandom Number Generators: Emerging Research and Opportunities.* IGI Global.

Bilan, S., Bilan, M., & Bilan, S. (2017). Research of the method of pseudo-random number generation based on asynchronous cellular automata with several active cells. *MATEC Web of Conferences, 125,* 1-6. 10.1051/matecconf/201712502018

Bilan, S., Bilan, M., Motornyuk, R., Bilan, A., & Bilan, S. (2016). Research and Analysis of the Pseudorandom Number Generators Implemented on Cellular Automata. *WSEAS Transactions on Systems, 15,* 275–281.

Chapter 6

Methods and Models for the Formation of Pseudo-Random Number Sequences Based on Cellular Automata

ABSTRACT

The chapter describes well-known models and implementation options for pseudorandom number generators based on cellular automata. Pseudorandom number generators based on synchronous and asynchronous cellular automata are briefly reviewed. Pseudorandom number generators based on one-dimensional and two-dimensional cellular automata, as well as using hybrid cellular automata, are described. New structures of pseudorandom number generators based on asynchronous cellular automata with a variable number of active cells are proposed. Testing of the proposed generators was carried out, which showed the high quality of the generators. Testing was conducted using graphical and statistical tests.

INTRODUCTION

The formation and obtaining of pseudorandom numbers is a necessary operation, which is widely used in various fields. The great need for pseudo-random number generators consists in the modeling of various dynamic processes. In other areas, the need for pseudo random number generators

DOI: 10.4018/978-1-7998-2649-1.ch006

(PRNG) has increased significantly. At the same time, there is a need for PRNG with specific properties for each problem. These properties satisfy his positive decision (Schneier, 1996; Marsaglia 2003). PRNG are implemented on the basis of various mathematical, hardware and software approaches. PRNG, which is implemented on the basis of cellular automata (CA), gained wide popularity and development. CA allow to implement a simple and understandable PRNG. Moreover, there are many paradigms that have a high quality pseudo-random bit sequence. Today, the properties of many CA have already been studied and methodological recommendations for building a PRNG based on CA of various configurations have been described (Bilan, 2017). The first generator that was implemented on the one-dimensional CA has been proposed by S. Wolfram (Wolfram, 1986). PRNGs, which are implemented on the hybrid CAs were considered in later works (Ruboi, et al 2004; Cattell, & Muzio, 1996). Hybrid cellular automata among a large number of identical cells contain cells that differ in their functional resources. Inhomogeneous cells can have different local logical functions, different neighborhoods, and other properties. The combination of rules for different CA cells are used in such generators. There were proposed some developed PRNGs, which are based on CA, that are implemented with usage of few CAs and additional generator. Such generators significantly improve the quality of work.. Additional generator is built on the linear feedback shift register (LFSR), as well as CA can be used to generate an additional sequence, for example, for LFSR bits (Suhinin, 2010a; Suhinin, 2010b; Hoe, et al 2012). Different approaches are used to analyze the properties of the PRNG. They are implemented as software products that are available in the Internet in appropriate sites (Walker, 2008; NIST Special Publications 800-22, 2001; NIST Special Publications 800-22, 2010; Marsaglia, 2003). They include such software as ENT, DIEHARD and NIST etc.

The paper analyzes three PRNGs based on the CA and there are the results of using NIST tests for analysis presented here. The more tests that are positive, the higher the quality of the generator. The most commonly used graphics tests are the DIEHARD and NIST tests.

This chapter discusses the use of ACA with a varying number of active cells for constructing pseudo-random number generators, with a large repetition period and high statistical properties.

PSEUDORANDOM NUMBER GENERATORS BASED ON ONE-DIMENSIONAL CELLULAR AUTOMATA

One-dimensional cellular automata are represented by one row or one column of cells. Such CA are called elementary cellular automata (ECA). Each cell contains a neighborhood of cells. A neighborhood is represented by a set of cells on both sides of each cell for which the neighborhood is considered (Wolfram, 1986a; Wolfram, 2002; Bilan, 2017). At each time step, each cell performs a certain logical function, the arguments of which are the signals of the state of the cells of the neighborhood. A cellular automaton with new conditions is being built. After a certain number of time steps, a two-dimensional picture is formed, which is called the evolution of the CA. The choice of LSF for each CA cell indicates the number of steps at which CA states do not repeat. If we take off the signals from the outputs of the cells and form a bit sequence from them, then we can talk about a pseudo-random bit sequence. From this point of view, a one-dimensional CA can be considered as a PRNG.

S. Wolfram considered the most detailed ECA. He published works in which he described elementary cellular automata and based on them formed the theory of true randomness (Wolfram, 1986a; Wolfram, 2002; Bilan, 2017). S. Wolfram describes a set of rules that each CA cell fulfills. A cell stets into a state that depends on its own state and on the states of neighboring cells at a previous point in time. The cell performs a function in accordance with a given rule and implements a result that indicates the state of the cell in the next time step.

S. Wolfram proposed the first PRNG based on ECA and also studied its properties in detail (Wolfram, 1986b). S. Wolfram explored many rules and identified generators that are similar to LFSR.

S. Wolfram showed that rule 30 generator gives a more random sequence (Wolfram, 1985; Wolfram, 1986; Wolfram, 1986b).

$$X_i(t+1) = X_{i-1}(t) \oplus \left[X_i(t) \vee X_{i+1}(t) \right].$$

PRNGs based on one-dimensional CA have been widely studied in (Cattel, et al 1999; Chowdhury, Nandy, & Chattopadhyay, 1994; Sipper, 1999; Sirakoulis, 2016; Cho, et al 2007). Much attention is paid to the study of ECA with various neighborhood structures.

However, PRNGs based on a one-dimensional CA did not prove high resistance to enemy attacks and cannot be used to protect information. They make it impossible to build a high-quality PRNG.

PSEUDORANDOM NUMBER GENERATORS BASED ON TWO-DIMENSIONAL CELLULAR AUTOMATA

PRNG Based on the SCA

A two-dimensional PRNG based on SCA is implemented by reading a bit signal from the output of one of the cells at each time step (Bilan, 2017). At each time step, all SCA cells perform LSF. The length of the repeat period of the PRNS based on the SCA depends on the selected LSF. The literature provides examples of the functioning of such a PRNG. They show that the initial state, LSF, and the shape of the neighborhood affect the duration of the period of operation of the generator. Moreover, such generators are an ordinary two-dimensional cellular automaton, which does not give high quality of the generated pseudo-random sequence of numbers. Using different LSFs does not significantly improve generator quality. To improve the operation of the PRNG, additional sources are used that pseudo-randomly change the state of the cellular automaton at each time step and extend the retry period. However, such generators are heterogeneous.

PRNG Based on the CA That the Additional Bits Using

Many papers describe the PRNG based on two-dimensional ACA (Bilan, 2017; Bilan, Bilan, & Bilan, 2015; Bilan, Bilan, & Bilan, 2017; Bilan, et al 2016). In such generators, the principle of operation is used, based on the implementation of LSF by only one active cell at a selected time step. However, such a PRNG has a short repeat period. Important is the choice of LTF and LSF.

PRNG is implemented in CA, in which only one cell (active cell) changes its own state at each time step. In such a CA, the cell neighborhoods are such that one active cell can be selected from the entire neighborhood of the active cell. This cell will be set to active.

The active cell selection model in the next time step is described in the previous sections. An activating signal can only be transmitted to one cell in this neighborhood. Each cell can contain the main (informational) and additional (active) states. A visual representation of active signal transmission and changes in the state of information on Figure 1 is presented.

Figure 1. The example of active signal transmission and the information state changes

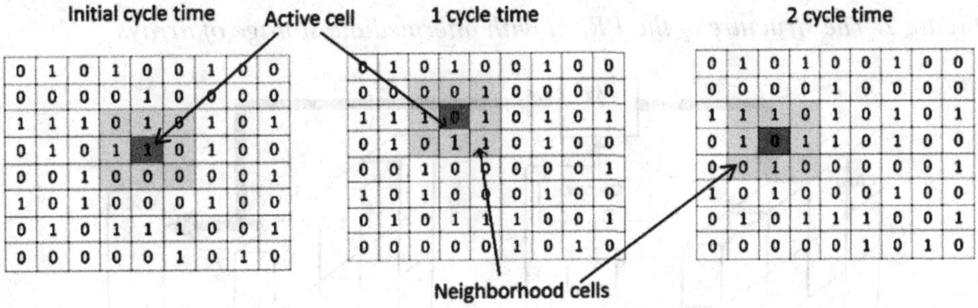

Pseudorandom bit sequences consist of bits generated at the outputs of active cells at each time step. At the same time, at each time step, the arrangement of active cells changes. That is, bits are take of from the outputs of different cells.

To achieve uniformity in the structure of the PRNG, additional sources are used as the basis for the functioning of the CA. Such sources can be any additional ACA cells, which at each time step are an additional cell of the neighborhood. They can change during the functioning of the CA.

The LSF arguments of each active cell are signals from the outputs of the cells of the neighborhood, its own output, and from the output of an additional cell, which is selected according to a predetermined law.

Each cell is connected with neighborhood cells by circuits of activation. The activating signal is transmitted by these circuits. The active cell analyzes the information state of neighborhood cells, forms the "1" signal on one of the active outputs, which is defined by a function of being activated and by the signals of neighborhood cells.

A PRNG is developed in which one active cell and one additional bit are used (Bilan, 2017; Bilan, Bilan, & Bilan, 2015; Bilan, et al 2016). The active signal is transmitted at even and odd time steps in different ways. At odd time steps of the generator's operation, a neighborhood cell is selected that is in logical "1" state and has the largest neighborhood number among the remaining neighborhood cells that are in logical "1" state. At an even time step, a cell of the neighborhood of the active cell is selected that has a logical state of "0" and has a higher numbering number among the other neighboring cells in the zero state.

PRNG consists of two CAs (Figure 2): main and additional (CA_{add}). CA_{add} is used to form an extra bit. One option is to use an additional CA that stores its state N × M time steps (where N × M is the size of a two-dimensional CA).

Figure 2. The structure of the PRNG with intermediate storage of arrays.

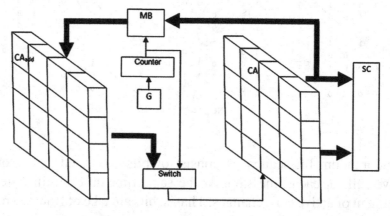

Additional bits are generated at the outputs of CAadd cells at each time step (Bilan, 2017; Bilan, Bilan, & Bilan, 2017; Bilan, et al 2016). To connect the outputs of CA_{add} cells, a switching system is used. An example of the operation of the generator on Figure 3 is presented.

Figure 3. An example of the generator functioning.

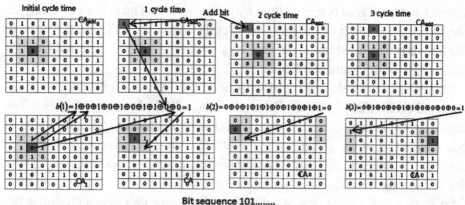

The second implementation option is PRNG, implemented on one CA (Bilan, 2017; Bilan, Bilan, & Bilan, 2017; Bilan, et al 2016). An additional bit is formed at the outputs of other cells of the same CA (for example, from left to right and from top to bottom). It is easier to implement and at the same time, at each time step, the state of the additional cell can be changed, which will be used further.

An example of the operation of the RNG on Figure 4 is presented.

Figure 4. Example of the PRNG with the inner CA scanning.

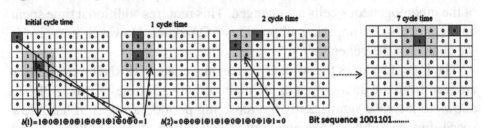

The high quality of such a PRNG depends on the well-chosen LTF and the law of the choice of additional bits. Additional bits should output the generator from the cycles and constantly affect the general state of the CA. Testing of such generators showed high quality of operation. Testing also showed that the used local transition function for active cells affects the quality of the generator.

PRNG Based on the CA With Heterogeneous Cells

The considered generators use several algorithms that are implemented for the ACA cell and for the formation of an additional bit. These generators uses a complicated switching circuitry for the constant switching the generator output to the data-output of active cells CA. Both generators have a large amount of a connections, which deteriorate the reliability of functioning.

The considered generators consist of homogeneous cells that realize the same LSF and LTF, and also have homogeneous bonds and connections with neighboring cells.

If you introduce in the ACA (replace cells with cells) such PRNG cells that realize another LSF, then ACA is called hybrid (Bilan, 2017; Ruboi, et al 2004). The laws of its functioning are changing. The repeat period of such a PRNG increases with an increase in the number of heterogeneous cells.

Test analysis of the operation of such PRNGs showed that the quality of work increases with an increase in the number of inhomogeneous cells. Hybrid PRNGs (HPRNG) based on ACA depend on the selected LSFs for the quality of work. Significantly improved quality if inhomogeneous cells realize different LSFs.

To increase the length of the repeat period of the bit pseudo-random sequence, a constant comparison of the hybrid ACA states at each time step is also used. If the states coincide at some time steps, then the coordinates of the inhomogeneous cells are changed. This requires additional time spent on the process of comparing all previous ACA states. However, it improves the quality of the generator.

The Figure 5 shows the work of orthogonal coated HPRNG. Formed bits are read from the data-output of the main cell at each timestep. This PRNG allows to increase significantly the repeating period, however, it has a low speed. Figure 6 shows an example of the operation of such a PRNG based on a cellular automaton with hexagonal coating.

Figure 5. An example of PRNG operation is based on the CA with heterogeneous cells and orthogonal coating HACA

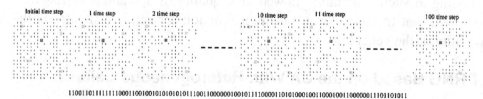

In this example, four heterogeneous cells are used that perform a majority function, while homogeneous cells perform the XOR function. The red cell

Figure 6. An example of PRNG operation is based on the CA with heterogeneous cells and hexagonal coating HACA

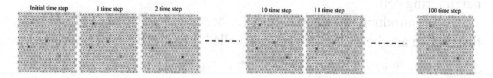

is the main one. A bit sequence is generated from its output, an example of which is presented in the Figure 6 for 100 time steps of the generator.

In this example, three heterogeneous cells are used. From the output of the red (main) cell, a pseudo-random bit sequence is formed. The initial settings for PRNG were random. If you use less than three inhomogeneous cells, the quality of such a generator deteriorates. The CA dimension also affects the quality of the generator.

PRNG BASED ON THE ACA WITH TWO ACTIVE CELLS

In an ACA with one active cell, minor changes are made at each time step. Only one cell can change state. All other cells are unchanged. Some ACA cells may remain unchanged during their entire functioning. Moreover, a generator based on such an ACA generates a high-quality pseudo-random bit sequence (Bilan, 2017; Bilan, et al 2016).

It is indisputable that the quality of the PRNG will be higher if a larger number of cells changes their state at each time step. If you use two active cells, both of these active cells will move along the ACA field. This eliminates the need for an extra bit. At each current time step, two ACA cells can simultaneously change their state and perform LSF. The quality of the generated bit sequence depends on the selected LTF for each active cell (Bilan, Bilan, & Bilan, 2017).

Figure 7 shows an example of PRNG based on ACA with two active cells.

Figure 7. An example of the work of PRNG based on the ACA with two active cells

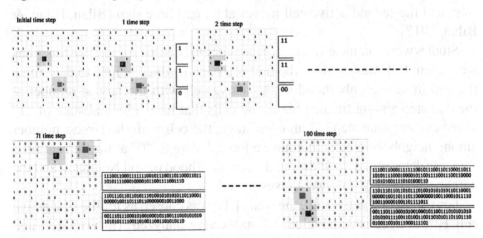

In the example shown in Figure 7, several ACA states are represented in the 100 time steps. Next to each time step are three bit sequences. The first bit sequence is formed by cells in the first active state. The second bit sequence is formed by cells in the second active state, and the third is the result of the XOR function of each bit of the first two sequences.

An ACA cell with two active states has more memory elements than an ACA cell with one active state. An additional flip-flop is used to memorize the second additional active state. Such a cell can work in four modes:

- Inactive state mode;
- First active state mode;
- Second active state mode;
- The mode of the first and second active state of the cell.

In the first mode, the cell does not perform LSF and LTF, but is in the standby mode of the transition signal to the active state. The cell has an informational state that was established at the initial moment of time or at previous time steps, when the cell was active. The cell is set into this mode at the next time step after an active state. In this mode, the cell can stay all the time ACA.

The modes of the first and second active states are described separately in previous sections.

The third mode is inherent in ACA in which two active cells have two different active states. In this mode, the cell performs one LSF and two LTF. In the case of using one active state, the coincidence of active cells leads to the loss of one active cell. To resolve this problem, the separation of time steps into even and odd is used. The first active cell moves at an even time step, and the second active cell moves at an odd time step (Bilan, Bilan, & Bilan, 2017).

Studies were conducted for such generators and two local transition functions were used. LTF was used for the first active cell, which helped to determine the cell in the neighborhood of the active cell with the largest number at the odd step among the neighborhood cells that have a logical state of "1" at the current time step. At an even stage, the cell with the largest number among neighboring cells that have a logical state of "0" at the current time step was determined. In addition, the neighborhood of von Neumann and the neighborhood of Moore were investigated.

For the second active cell, the same LTF was chosen as for the first active cell. However, cells with logical "1" states were analyzed at an even step, and at an odd step all cells of the neighborhood of the zero-state were analyzed, and the cell with the largest numbering among the neighborhood cells was determined.

The LTFs of the first and second active cells are realized identically, but they analyze cells with opposite binary states.

PRNG based on ACA with two active cells forms three bit sequences Q_1, Q_2, $Q_3 = Q_1 \oplus Q_2$. Each bit at the output of active cells is formed using the XOR function over the signals of the neighborhood cells, the eigenstate and the state of the additional cell at the current time step. An additional bit can be generated using the method presented in the works (Bilan, 2017; Bilan, Bilan, & Bilan, 2015; Bilan, et al 2016).

The results of testing all three sequences generated by PRNG using graphical tests on Figure 8 and Figure 9 are shown.

To perform the test the long sequences that have 2000000 bit in each sequence were formed. Graphical tests show bursts of amplitudes for bit sequences of each active cell. However, the third bit sequence Q_3 shows good test results. At the same time, only two LTFs described earlier were realized and other local transition functions were not analyzed.

Figure 8. The results of testing of all three of bit sequences with help a graphical test of the distribution of bit sequence elements on the plane

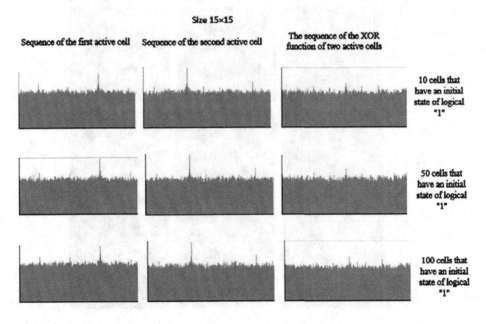

ACA in various sizes were used. For large ACA dimensions, the tests showed good results. Good results were also obtained for small sizes, which was not observed for ACA with one active cell.

PRNG Based on ACA With a Variable Number of Active Cells

A generator with two active cells is used. Active cells perform the same function of the local state at each time step. If at some point in time two active cells are combined into one ACA cell, then only one cell changes its

Figure 9. Diagrams of the distribution of points in the plane for bit sequences formed using the Moore neighborhood for a sequence of length of 2,000,000 bits

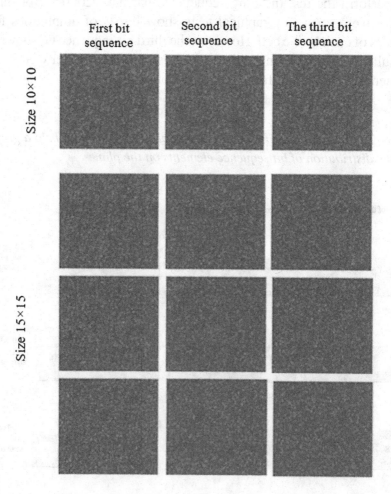

state. When both active cells are combined in one ACA cell, they form a new active cell. This new active cell in the next time step is a cell that is located at the site of the union of two active cells. She appears in the next time step.

Thus, in the next time step, after combining two active cells in one ACA cell, three active cells appear. One of them is new and is implementing a new LTF. In the next time step, the two original active cells go to other cells, and the third active cell goes to another active cell in accordance with the set LTF. From this moment, already three AKA cells can change their state. In this case, the cell cannot change its state if the result of the execution of the local state function matches the value in the previous step. An example of the functioning of the original two active cells and the new active cell on Figure 10 is shown.

Figure 10. An example of the functioning of ACA at the time of the formation of a new active cell in ACA a well as at the following time steps

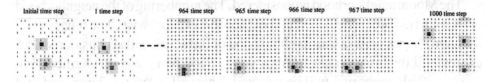

In Figure 10, the first active cells are green and red. The cells neighborhood each active cell are highlighted in yellow. At time step 965, two active cells coincide in one cell. At 966 time step, a new active blue cell appears. In the subsequent steps of evolution, three active cells are visible. It was initially proposed that in the ACA field there can be only three active cells. No more cells forming. The work consider the mode of operation of ACA, when new active cells are formed only by the initial two active cells. There may be other modes for the formation of new active cells in the ACA. For example, new active cells can also form new active cells in future. In addition, new active cells can form under the condition that two active cells are combined with certain LTFs. Thus, after a certain number of time steps, all cells can change their state.

According to the new LTF at the next time step, the active cell is determined by the code of the first three cells of the neighborhood of the third active cell at the current time step. Numbering for active cells of the Moor neighborhood on Figure 11 is shown.

Figure 11. An example of the numbering of cells of the Moore neighborhood, which was used in the software implementation of the method for the first, second and third active cells

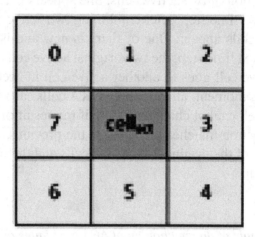

The Moore neighborhood is considered. The numbering of the neighborhood cells is carried out from the left cell of the top row clockwise. The binary code of the first three cells in the neighborhood of the active cell is considered. For example, if the binary code corresponds to 010, then the cell at number 3 in the specified numbering will be selected. This cell of the neighborhood will go into an active state at the next time step. An example of the transmission of an active signal by a third active cell on Figure 12 is shown.

Figure 12. An example of LTF performing by a third active cell

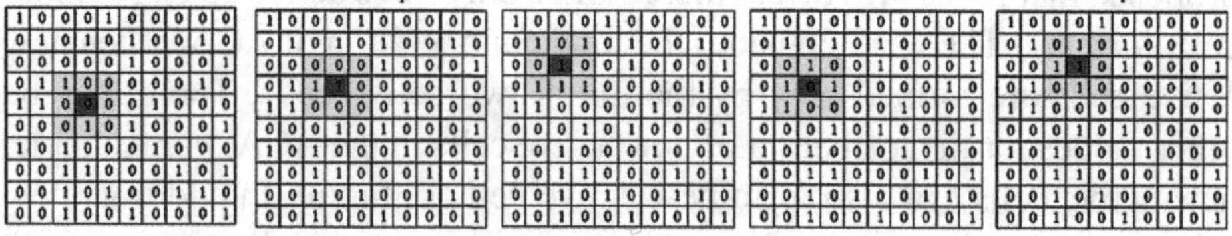

1 time step	2 time step	3 time step	4 time step	5 time step

In this paper, an example is considered where the first and second active cells perform LTF, which are described in the previous section. These cells have a neighborhood where all neighborhood cells are numbered according to Figure 11. For such LTF the cells can only be combined in the case shown in Figure 13.

Figure 13. The combination of cells of the neighborhoods of two active cells, which may lead to their coincidence

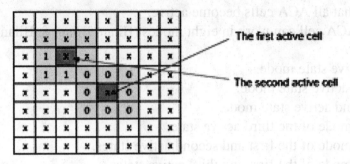

At the next time step, the cells are combined (Figure 14). At this time step, a new cell is formed, and at the next time step, the first and second active cells move to other positions according to the established LTFs. The new active cell is blue and moves according to the previously described LTF. She also begins to move to next active cell on the next time step. For other LTFs, the number of variants preceding the overlap may be greater. For example, for a new cell, the described LTF allows to increase the number of possible combinations for combining.

Figure 14. Example of the formation of a new active cell and their transition according to the given LTF

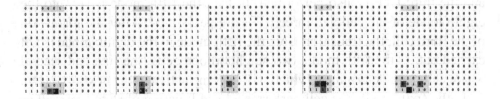

The symbols "x" indicate that the state can be any in these cells (0 or 1). At the following time steps, in some cells, the states were chosen arbitrarily. This was necessary to select the following active cells that belong to the neighborhood of the active cells. It is shown how transitions of active cells are carried out.

The use of different LTFs for each active cell does not allow the active states of the cell to be removed when their location coincides. However, it is possible that all ACA cells become active.

Each ACA cell can work in eight modes (Bilan, Bilan, & Bilan, 2017).

1.   Inactive state mode.
2.   First active state mode.
3.   Second active state mode.
4.   The mode of the third active state.
5.   The mode of the first and second active states.
6.   The mode of the first and third active states.
7.   The mode of the second and third active states.
8.   The mode of the first, second and third active states.

In the first mode, the ACA cell is not active and does not perform local transition functions and a local state function. In this mode, the cell does not change its state. The cell is in this mode at the beginning of the ACA or after it was in an active state.

If the cell set into the second, third or fourth modes, then it performs a local state function and the corresponding LTF for the first, second or third active cell. In our example, all active cells perform the same LSF and different LTF.

In the fifth mode, the cell performs a local function of states and two LTFs that are installed for the first two active cells. Also in this mode, the cell forms the third active cell (if the fifth mode has occurred for the first time), and the third LTF is also performed. In fact, the fifth mode is set into the eighth mode of the cell and three LTFs are performed. After the formation of the third cell, the transition from the fifth mode to the eighth mode is not carried out. In the fifth mode, two neighborhood cells (the first and second active cells) are determined, which will become active in the next time step.

In the sixth mode, the cell performs a local function of states, as well as the first and third LTF. In the seventh mode, the cell performs a local function of states, as well as the second and third LTF.

In the eighth mode, the cell performs a local function of states, as well as the first, second, and third LTF.

The paper considers the option with only three active cells. PRNG quality analysis is performed using graphical tests. Results were evaluated for all generated bit sequences by all active cells and sequences generated using XOR functions over the bits of active cell sequences. Figure 15 shows the results of tests of the distribution of numbers on the plane for all sequences.

Figure 15. Results of using a graphic test for the distribution of elements of bit sequences on a plane

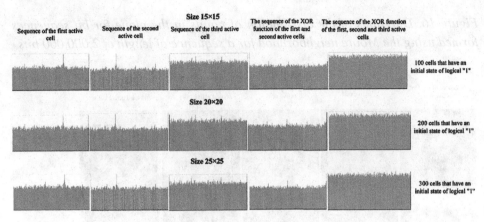

The generated bit sequences for each initial state of the generator were analyzed. Bit sequences of 2,000,000 bits were used. This allows to clearly formulate the results of graphic tests. For each initial state, the corresponding number of cells set to the logical "1" state was used. The quality of the generated bit sequence depends on this amount. For each initial state of the ACA, five bit sequences were formed and their quality was evaluated using graphic tests. The first three bit sequences were formed by the first, second, and third active cells. The fourth bit sequence is obtained by performing the XOR function for all bits of the first and second bit sequences that are formed by the first and second active cells. The fifth bit sequence is obtained by performing the XOR function for all bits of the first and second bit sequences that are formed by the first, second and third active cells. Moreover, the third bit sequence is always less than the first two sequences since it begins to form later after the first coincidence of the first two active cells. The greatest length of the third bit sequence can be obtained if the initial state of the first two active cells is used. The results are presented for the sizes of ACA 15 × 15, 20 × 20 and 25 × 25 cells, as well as for a different number of cells, which at the initial moment have a state of logical "1".

The obtained histograms show that the LTF of the third cell gives a better sequence, and the fourth and fifth resulting bit sequences have a good distribution.

The quality of the obtained bit sequences is also confirmed by the graphic test of the distribution of numbers on the plane as shown in Figure 16. The results for the same bit sequences are presented (Figure 15).

115

This graphical test showed good results for all bit sequences.

Figure 16. Diagrams of the distribution of points in the plane for bit sequences formed using the Moore neighborhood for a sequence of length of 2,000,000 bits

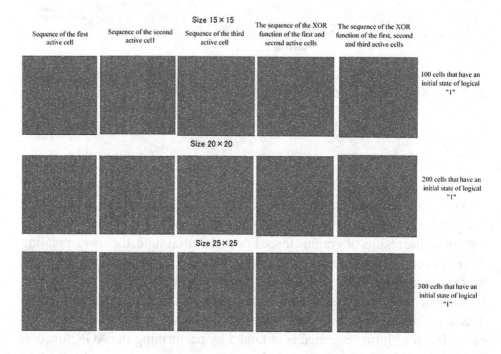

Figure 17 shows the evolution of an ACA based on a hexagonal coating. Two active cells are used. In the ninth time step, the cells form a new active cell, which is determined by the new LTF.

Figure 17. An example of the functioning of a PRNG with a varying number of active cells based on an ACA with a hexagonal coating

The HexRND3 program is written in the programming language C# simulates the operation of a cellular automaton with hexagonal coating, designed to generate a random sequence of bits and at the initial moment of time containing two active cells. Moreover, in the process of the working cycle, provided that these two active cells meet, a third appears. Each active cell has its own local function and transition rule. The local function may be one of 5:

1. $x0 = x0 = x0 \wedge x1 \wedge x2 \wedge x3 \wedge x4 \wedge x5 \wedge x6$
2. $x0 = x0 * x1 * x2 * x3 * x4 * x5 * x6$
3. $x0 = x1*x2 + x3*x4 + x5*x6$
4. $x0 = (x1+x2) \wedge (x3+x4) \wedge (x5+x6)$
5. $x0 = (x1+!x2) \wedge (x3+!x4) \wedge (x5+!x6)$

were $x0$ – cell value; $x1, x2, x3, x4, x5, x6$ – the value of neighboring cells.

A fragment of the source code that implements the desired local function of the active cell has the form of a single operator in the Appendix is presented:

In addition, at each step of the iteration, one additional bit is determined, the value of which is equal at the first step to the value of the first CA cell, at the second - to the value of the second cell, and so on along the entire plane of the automaton. When the last cell is reached, the active bit goes over again to the first. When the operating mode is activated, taking into account this additional bit, the local functions presented above take the form:

1. $x0 = x0 \wedge x1 \wedge x2 \wedge x3 \wedge x4 \wedge x5 \wedge x6 \wedge xa$
2. $x0 = x0 * x1 * x2 * x3 * x4 * x5 * x6 \wedge xa$
3. $x0 = x1*x2 + x3*x4 + x5*x6 \wedge xa$
4. $x0 = (x1+x2) \wedge (x3+x4) \wedge (x5+x6) \wedge xa$
5. $x0 = (x1+!x2) \wedge (x3+!x4) \wedge (x5+!x6) \wedge xa$

were $x0$ – cell value; $x1, x2, x3, x4, x5, x6$ – the value of neighboring cells, xa – value of the additional bit.

The source code for calling a miscalculation of a local function, taking into account the additional bit:

```
KA[active_cell[i]] = _get_bit(active_cell[i], active_func[i]) ^
(checkBox1.Checked & addBit);
```

There are 7 rules for active cell transitions:

1 - "senior of the majority (h.p.)"
2 - "even code (0.2.4)"
3 - "odd code (1,3,5)"
4 - "code in junior (0,1,2)"
5 - "code in the older (3,4,5)"
6 - "senior '1' (by h.p.)"
7 - "senior '0' (by h.p.)"

A fragment of the source code for determining the transition of the active cell in the Appendix is presented.

One of the unique features is that in fact a two-dimensional matrix of a CA with a hexagonal coating is programmatically represented by a one-dimensional array. For this purpose, a technique was specially developed for converting the index index number of this array to the specific location of the corresponding cell on the plane. In the Appendix, for example, is a fragment of the source code for initialization of a one-dimensional array and its visualization using controls label depending on the sizes (in X and Y) of the simulated CA is presented.

A certain difficulty in using such a one-dimensional array to represent a hexagonal CA arises when determining neighbors. Moreover, in contrast to the search for neighboring cells, aggravated only by shifting the odd rows (as is the case with the HexRND2 program), this must be done, knowing only one serial number - the index. The function that returns the indices of 6 neighboring cells by one index of the current element in the Appendix is represented by the source code.

Simulation can be stopped and the current state of the CA saved to an external file. Source code for export implementation in the Appendix is described.

The CA's previously saved state can be imported from an external file into the program to continue modeling. Import implementation source code in the Appendix is presented.

The modeling program is based on a miscalculation of one iteration of the CA, performed according to the following algorithm:

1. Definitions of the value of the additional bit.
2. Calculation of transition functions of active cells.
3. If the number of active cells = 2 and they did not meet, go to step 6.
4. If the number of active cells = 3 - go to step 6.

Figure 18. Graphical test charts for PRNG based on a hexagonal coated ACA

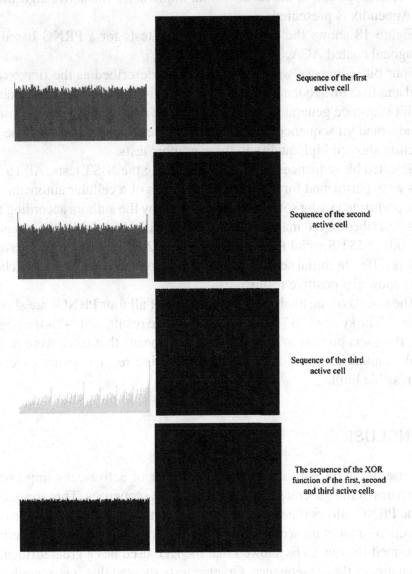

Sequence of the first active cell

Sequence of the second active cell

Sequence of the third active cell

The sequence of the XOR function of the first, second and third active cells

5. A third active cell is added, for which its own local function and transition rule are assigned.
6. Calculation of local functions of all active cells in accordance with the setting of the additional bit.
7. CA generates an output bit of the sequence, which is equal to the XOR function over all active cells of the current iteration. Go to step 1.

The source code of the function that implements the above algorithm on the Appendix is presented.

Figure 18 shows the results of graphical tests for a PRNG based on a hexagonal coated ACA.

Four bit sequences were used, which are described as the first, second, third and fifth bit sequences described in Figures 63 and 64. In this example, the bit sequence generated by the third active cell is unsuitable for use as a pseudo-random sequence. However, the sequence combined with the XOR function showed high quality in these graphic tests.

Selected bit sequences were evaluated using the NIST tests. All 16 NIST tests were performed for different initial states of a cellular automata. Tests were performed using software implemented by the authors according to the tests described in the manual (Bilan, 2017; NIST Special Publications 800-22, 2001; NIST Special Publications 800-22, 2010). Different ACA sizes as well as different initial settings of active cells on the AKA were also chosen. Tests showed a positive result.

The results of the analyzing the behavior of all four PRNGs are shown in Table 1. The symbol "+" indicates the positive result, and "-" – the negative one. If «±» is present in the table cell, this means that there were negative results among positive ones but these negative results do not exceed the permissible limits.

CONCLUSION

The use of ACAs with an increasing number of active cells improves the properties of the generated pseudo-random bit sequences. This arrangement of the PRNG allows the use of additional internal sources for the formation of pseudo-random bit sequences. The analysis of ACA with active cells that performed various LTFs showed that the LTF used has a great influence on the quality of the bit sequence. Graphic tests showed that it is enough to use the bit sequence of one active cell with the best statistical characteristics. This proves the impact of LTF on the performance of a generator based on ACA. Also, graphical tests showed that the Moore neighborhood is the most effective. It gives a positive result for small ACA sizes when using graphical tests. The use of the XOR function for bit sequences formed by two and three active cells gives the best results, as proved by graphic tests. The use of additional active cells increases the length of the repeat period of the pseudo-random bit sequence.

Table 1. Results of the analysis of the PRNG (Only the Moore neighborhood)

N° of Test	ACA Size	The Initial Number of Cells With a Logical State of "1"	The Number of Active Cells	The Length of the Bit Sequence	Test Result				
					Output of the First Active Cell	Output of the Second Active Cell	Output of the third Active +Cell	XOR Function of the First and Second Active Cells	XOR Function of the First, Second and Third Active Cells
1	25×25	100, 315	7	1000000	+	+	+	+	+
	30×30	190	5	1000000	+	+	+	+	+
	30×30	770	3	1000000	+	+	+	+	+
2	25×25	100, 315	7	10000	+	+	+	+	+
	30×30	190	5	1000	+	+	+	+	+
	30×30	770	3	10000	+	+	+	+	+
3	25×25	100, 315	7	10000	+	+	+	+	+
	30×30	190	5	1000	+	+	+	+	+
	30×30	770	3	100000	+	+	+	+	+
4	25×25	100, 315	7	6272	+	+	+	+	+
	30×30	190	5	6272	+	+	+	+	+
	30×30	770	3	6272	+	+	+	+	+
5	25×25	100, 315	7	1000	+	+	+	+	+
	30×30	190	5	1000	+	+	+	+	+
	30×30	770	3	1000, 2000	+	+	+	+	+
6	25×25	100, 315	7	10000	+	+	+	+	+
	30×30	190	5	10000	+	+	+	+	+
	30×30	770	3	100000	+	+	+	+	+
7	25×25	100, 315	7	10000	+	+	+	+	+
	30×30	190	5	100000	+	+	+	+	+
	30×30	770	3	10000, 100000	+	+	+	+	+
8	25×25	100, 315	7	10000	+	+	+	+	+
	30×30	190	5	10000	+	+	+	+	+
	30×30	770	3	10000	+	+	+	+	+
9	25×25	100, 315	7	387840	+	+	+	+	+
	30×30	190	5	387840	+	+	+	+	+
	30×30	770	3	387840	+	+	+	+	+
10	25×25	100, 315	7	10000	+	+	+	+	+
	30×30	190	5	100000	+	+	+	+	+
	30×30	770	3	1000000	+	+	+	+	+
11	25×25	100, 315	7	1200000	+	+	+	+	+
	30×30	190	5	1000000	+	+	+	+	+
	30×30	770	3	1000000	+	+	+	+	+
12	25×25	100, 315	7	10000	+	+	+	+	+
	30×30	190	5	100000	+	+	+	+	+
	30×30	770	3	10000	+	+	+	+	+
13	25×25	100, 315	7	10000	+	+	+	+	+
	30×30	190	5	10000	+	+	+	+	+
	30×30	770	3	10000	+	+	+	+	+
14	25×25	100, 315	7	10000	+	+	+	+	+
	30×30	190	5	1000	+	+	+	+	+
	30×30	770	3	10000	+	+	+	+	+
15	25×25	100, 315	7	1000000	+	+	+	+	+
	30×30	190	5	1000000	+	+	+	+	+
	30×30	770	3	1000000	+	+	+	+	+
16	25×25	100, 315	7	1000000	+	+	+	+	+
	30×30	190	5	1000000	+	+	+	+	+
	30×30	770	3	1000000	+	+	+	+	+

NIST test results show that the generator passed all the tests successfully, if you use the XOR function for bit sequences generated at the outputs of active cells.

REFERENCES

Bilan, S. (2017). *Formation Methods, Models, and Hardware Implementation of Pseudorandom Number Generators: Emerging Research and Opportunities.* IGI Global.

Bilan, S., Bilan, M., & Bilan, S. (2015). Novel pseudorandom sequence of numbers generator based cellular automata. *Information Technology and Security, 3*(1), 38–50.

Bilan, S., Bilan, M., & Bilan, S. (2017). Research of the method of pseudo-random number generation based on asynchronous cellular automata with several active cells. *MATEC Web of Conferences, 125*, 1-6. 10.1051/matecconf/201712502018

Bilan, S., Bilan, M., Motornyuk, R., Bilan, A., & Bilan, S. (2016). Research and Analysis of the Pseudorandom Number Generators Implemented on Cellular Automata. *WSEAS Transactions on Systems, 15*, 275–281.

Cattel, K., Zang, S., Serra, M., & Muzio, J. C. (1999). 2-by-n hybrid cellular automata with regular configuration: Theory and application. *IEEE Transactions on Computers, 48*(3), 285–295. doi:10.1109/12.754995

Cattell, K., & Muzio, J. C. (1996). Synthesis of one-dimensional linear hybrid cellular automata. *IEEE Trans. on Computer-Aided Design of Integrated Circuits and Systems, 15*(3), 325-335.

Cho, S.J., Choi, U. S., Kim, H.D., Hwang, Y.H., Kim, J.G., & Heo, S.H. (2007). New synthesis of one-dimensional 90/150 liner hybrid group CA. *IEEE Transactions on Computer-Aided Design of Integrated Circuits and Systems, 25*(9), 1720-1724.

Chowdhury, D. R., Nandy, S., & Chattopadhyay, S. (1994). Additive cellular automata. *Theory and Applications Journal, 1*(2), 12–15.

Hoe, D. H. K., Comer, J. M., Cerda, J. C., Martinez, C. D., & Shirvaikar, M. V. (2012). Cellular Automata-Based Parallel Random Number Generators Using FPGAs. *International Journal of Reconfigurable Computing, 2012*, 1–13. doi:10.1155/2012/219028

Marsaglia, G. (2003). Random number generators. *Journal of Modern Applied Statistical Methods; JMASM, 2*(1), 2–13. doi:10.22237/jmasm/1051747320

NIST Special Publications 800-22. (2001). *A statistical test suite for random and pseudorandom number generators for cryptographic applications.* NIST.

NIST Special Publications 800-22. (2010). A statistical test suite for random and pseudorandom number generators for cryptographic applications. Revision 1. NIST. https://csrc.nist.gov/groups/ST/toolkit/rng/documentation_software.html

Ruboi, C. F., Hernandez Encinas, L., White, S. H., del Rey, A. M., & Sancher, R. (2004). The use of Linear Hybrid Cellular Automata as Pseudorandom bit Generators in Cryptography. *Neural Parallel & Scientific Comp.*, *12*(2), 175–192.

Schneier, B. (1996). Applied Cryptography, Second Edition: Protocols, Algorthms, and Source Code in C. Wiley Computer Publishing, John Wiley & Sons, Inc.

Sipper, M. (1999). The emergence of cellular computing. *Computers*, *32*(7), 18–26. doi:10.1109/2.774914

Sirakoulis, G. C. (2016). Parallel Application of Hybrid DNA Cellular Automata for Pseudorandom Number Generation. *JCA*, *11*(1), 63–89.

Suhinin, B.M. (2010a). High generators of pseudorandom sequences based on cellular automata. *Applied Discrete Mathematics*, *2*, 34 – 41.

Suhinin, B. M. (2010b). Development of generators of pseudorandom binary sequences based on cellular automata. *Science and education*, (9), 1–21.

Walker, J. (2008). *ENT. A Pseudorandom Number Sequence Test Program.* http://www.fourmilab.ch/random

Wolfram, S. (1985). Origins of randomness in physical system. *Physical Review Letters*, *55*(5), 449–452. doi:10.1103/PhysRevLett.55.449 PMID:10032356

Wolfram, S. (1986). Cryptography with Cellular Automata. *Lecture Notes in Computer Science*, *218*, 429–432. doi:10.1007/3-540-39799-X_32

Wolfram, S. (1986a). *Appendix of Theory and Applications of Cellular Automata.* World Scientific.

Wolfram, S. (1986b). Random Sequence Generation by Cellular Automata. *Advances in Applied Mathematics*, 7(2), 429–432. doi:10.1016/0196-8858(86)90028-X

Wolfram, S. (2002). *A new kind of science*. Wolfram Media.

APPENDIX

The HexRND3 program is written in the programming language C# simulates the operation of a cellular automaton with hexagonal coating, designed to generate a random sequence of bits and at the initial moment of time containing two active cells. A fragments of the source code on The Appendix is presented. A fragment of the source code that implements the desired local function of the active cell has the form of a single operator in the Appendix is presented. A fragment of the source code for determining the transition of the active cell the Appendix is presented. For example, is a fragment of the source code for initialization of a one-dimensional array and its visualization using controls label depending on the sizes (in X and Y) of the simulated CA is presented. Source code for export implementation in the Appendix is described.

A fragment of the source code that implements the desired local function of the active cell has the form of a single operator:

```
switch (func)
{
    case 1:
        ret = KA[index_pe] ^ KA[s[0]] ^ KA[s[1]] ^ KA[s[2]] ^
KA[s[3]] ^ KA[s[4]] ^ KA[s[5]];
        break;
    case 2:
        ret = KA[index_pe] && KA[s[0]] && KA[s[1]] && KA[s[2]]
&& KA[s[3]] && KA[s[4]] && KA[s[5]];
        break;
    case 3:
        ret = (KA[s[0]] & KA[s[1]]) | (KA[s[2]] & KA[s[3]]) |
(KA[s[4]] & KA[s[5]]);
        break;
    case 4:
        ret = (KA[s[0]] | KA[s[1]]) ^ (KA[s[2]] | KA[s[3]]) ^
(KA[s[4]] | KA[s[5]]);
        break;
    case 5:
        ret = (KA[s[0]] | !KA[s[1]]) ^ (KA[s[2]] | !KA[s[3]]) |
(KA[s[4]] ^ !KA[s[5]]);
        break;
    default:
        ret = false;
        break;
}
```

A fragment of the source code for determining the transition of the active cell:

```
public int _get_perehod(int index_pe, int func)
        {
            int ret = index_pe;
            int[] s = _get_KA_sosedi(index_pe);
            switch (func)
            {
                case 1:
                    int kol1 = Convert.ToInt16(KA[index_pe]) +
Convert.ToInt16(KA[s[0]]) +
                            Convert.ToInt16(KA[s[1]]) +
Convert.ToInt16(KA[s[2]]) +
                            Convert.ToInt16(KA[s[3]]) +
Convert.ToInt16(KA[s[4]]) + Convert.ToInt16(KA[s[5]]);
                    Boolean m1 = (kol1 > 3);
                    for (int i = 5; i > 0; i--)
                    {
                        if (!(KA[s[i]] ^ m1))
                        {
                            ret = s[i];
                            break;
                        }
                    }
                    break;
                case 2:
                    ret = s[_bits2num(KA[s[0]], KA[s[2]],
KA[s[4]])];
                    break;
                case 3:
                    ret = s[_bits2num(KA[s[1]], KA[s[3]],
KA[s[5]])];
                    break;
                case 4:
                    ret = s[_bits2num(KA[s[0]], KA[s[1]],
KA[s[2]])];
                    break;
                case 5:
                    ret = s[_bits2num(KA[s[3]], KA[s[4]],
KA[s[5]])];
                    break;
                case 6:
                    for (int i = 5; i > 0; i--)
                    {
                        if (KA[s[i]])
                        {
                            ret = s[i];
```

```
                                  break;
                          }
                  }
                  break;
              case 7:
                  for (int i = 5; i > 0; i--)
                  {
                          if (!(KA[s[i]]))
                          {
                                  ret = s[i];
                                  break;
                          }
                  }
                  break;
              default:
                  ret = s[0];
                  break;
          }
          return ret;
      }
      private int _bits2num(Boolean b0, Boolean b1, Boolean
b2)
      {
          int num = -1;
          if (b0)
          {
                  num = num + 1;
          }
          if (b1)
          {
                  num = num + 2;
          }
          if (b2)
          {
                  num = num + 4;
          }
          if (num > 5)
          {
                  num = 5;
          }
          if (num < 0)
          {
                  num = 0;
          }
          return num;
      }
```

The fragment of the source code for initialization of a one-dimensional array and its visualization using controls label depending on the sizes (in X and Y) of the simulated CA.

```
public void init_KA(Boolean KA_from_file)
{
    lenKAx = Convert.ToInt32(comboBox3.SelectedItem);
    lenKAy = Convert.ToInt32(comboBox4.SelectedItem);
    lenKA = lenKAx * lenKAy;
    cell_size = Convert.ToInt32(comboBox5.SelectedItem);
    panel1.Visible = false;
    panel1.Controls.Clear();
    panel1.Visible = true;
    int linKA_plus_1 = lenKA + 1;
    if (!KA_from_file)
    {
        KA = new Boolean[linKA_plus_1];
        Array.Clear(active_cell, 0, 3);
        Array.Clear(active_func, 0, 3);
        Array.Clear(active_perehod, 0, 3);
    }
    active1_bytes.Clear();
    active2_bytes.Clear();
    active3_bytes.Clear();
    RND_bytes.Clear();
    tb = new Label[linKA_plus_1];
    bits_counter = 0;
    panel1.Width = lenKAx * (cell_size + 1) + cell_size / 2 +
4;
    panel1.Height = lenKAy * (cell_size + 1) + 6;
    label1.Visible = true;
    label1.BringToFront();
    label1.Invalidate();
    panel1.Visible = false;
    panel1.Enabled = false;
    panel1.Invalidate();
    for (int y = 0; y < lenKAy; y++)
    {
        for (int x = 0; x < lenKAx; x++)
        {
            int index = y * lenKAx + x + 1;
            tb[index] = new System.Windows.Forms.Label();
            tb[index].Location = new System.Drawing.Point(2 + x
* (cell_size + 1) + (y % 2) * (cell_size / 2), 2 + y * (cell_
size + 1));
            tb[index].Name = "label_" + (index + 1000).
ToString();
            tb[index].AutoSize = false;
```

```
            tb[index].Size = new System.Drawing.Size(cell_size,
cell_size);
            tb[index].MinimumSize = new System.Drawing.
Size(cell_size, cell_size);
            tb[index].Font = new Font("Courier New", 10);
            tb[index].BorderStyle = BorderStyle.FixedSingle;
            tb[index].BackColor = Color.White;
            tb[index].Cursor = Cursors.Hand;
            tb[index].TextAlign = ContentAlignment.
MiddleCenter;
            toolTip1.SetToolTip(this.tb[index], index.
ToString());
            tb[index].MouseClick += new MouseEventHandler(tb_
MouseClick);
            tb[index].MouseDoubleClick += new
MouseEventHandler(tb_MouseDoubleClick);
            tb[index].Visible = false;
            panel1.Controls.Add(tb[index]);
        }
    }
    for (int i = 1; i <= lenKA; i++)
    {
        tb[i].Visible = true;
    }
    label1.Visible = false;
    label1.Invalidate();
    panel1.Visible = true;
    panel1.Enabled = true;
    panel1.Invalidate();
    if (!KA_from_file)
    {
        if (radioButton1.Checked)
        {
            _KA_0fill();
        }
        else if (radioButton2.Checked)
        {
            _KA_1fill();
        }
        else if (radioButton3.Checked)
        {
            _KA_randomfill();
        }
    }
    else
    {
        KA_2_tb();
    }
```

```
    _clear_all_results();
}
```

The function that returns the indices of 6 neighboring cells by one index of the current element

```
public int[] _get_KA_sosedi(int index)
{
    int[] s = new int [6] {0, 0, 0, 0, 0, 0};
    int ostatok_x = index % lenKAx;
    int chet_y;
    if (ostatok_x == 0)
    {
        chet_y = (index / lenKAx) % 2;
    }
    else
    {
        chet_y = ((index / lenKAx) + 1) % 2;
    }
    Boolean gran_U, gran_D, gran_L, gran_R;
    gran_U = (index <= lenKAx);
    gran_D = (index > (lenKA - lenKAx));
    gran_L = (ostatok_x == 1);
    gran_R = (ostatok_x == 0);
    s[0] = index - lenKAx - chet_y;
    s[1] = index - lenKAx + 1 - chet_y;
    s[2] = index + 1;
    s[3] = index + lenKAx + 1 - chet_y;
    s[4] = index + lenKAx - chet_y;
    s[5] = index - 1;
    if (gran_U)
    {
        s[0] = s[0] + lenKA;
        s[1] = s[1] + lenKA;
    }
    if (gran_D)
    {
        s[3] = s[3] - lenKA;
        s[4] = s[4] - lenKA;
    }
    if (gran_L)
    {
        s[5] = s[5] + lenKAx;
        if (chet_y == 1)
        {
            s[0] = s[0] + lenKAx;
            s[4] = s[4] + lenKAx;
        }
```

```
    }
    if (gran_R)
    {
        s[2] = s[2] - lenKAx;
        if (chet_y == 0)
        {
            s[1] = s[1] - lenKAx;
            s[3] = s[3] - lenKAx;
        }
    }
    for (int i = 0; i < 6; i++)
    {
        if (s[i] == 0)
        {
            MessageBox.Show("Neighbor s[" + i.ToString() +
"]");
        }
    }
    return s;
}
```

Source code for export implementation.

```
saveFileDialog1.InitialDirectory = Application.StartupPath +
"\\";
saveFileDialog1.ShowDialog();
if (saveFileDialog1.FileName != "")
{
    file_KA_export = new System.
IO.StreamWriter(saveFileDialog1.FileName, false, Encoding.
GetEncoding(1251));
    file_KA_export.AutoFlush = true;
    for (int i = 0; i < 3; i++)
    {
        file_KA_export.WriteLine(active_cell[i].ToString() +
";" + active_func[i].ToString() + ";" + active_perehod[i].
ToString());
    }
    file_KA_export.WriteLine(lenKAx.ToString() + ";" + lenKAy.
ToString());
    foreach (Boolean bit in KA)
    {
        file_KA_export.Write((Convert.ToInt16(bit)).
ToString());
    }
    file_KA_export.Close();
    file_KA_export.Dispose();
}
```

Import implementation source code.

```
StreamReader myStream;
openFileDialog1.InitialDirectory = Application.StartupPath +
"\\";
if (openFileDialog1.ShowDialog() == DialogResult.OK)
{
    try
    {
        myStream = new System.IO.StreamReader(openFileDialog1.
FileName);
        string[] s3 = new string[3];
        Array.Clear(active_cell, 0, 3);
        Array.Clear(active_func, 0, 3);
        Array.Clear(active_perehod, 0, 3);
        for (int i = 0; i < 3; i++)
        {
            s3 = myStream.ReadLine().Split(';');
            active_cell[i] = Convert.ToInt32(s3[0]);
            active_func[i] = Convert.ToInt32(s3[1]);
            active_perehod[i] = Convert.ToInt32(s3[2]);
        }
        s3 = myStream.ReadLine().Split(';');
        lenKAx = Convert.ToInt32(s3[0]);
        lenKAy = Convert.ToInt32(s3[1]);
        lenKA = lenKAx * lenKAy;
        comboBox3.SelectedItem = lenKAx.ToString();
        comboBox4.SelectedItem = lenKAy.ToString();
        KA = new Boolean[lenKA + 1];
        int pe;
        for (int i = 0; i <= lenKA; i++)
        {
            KA[i] = (myStream.Read() == 49);
        }
        init_KA(true);
    }
    catch (Exception ex)
    {
        MessageBox.Show("Error of reading file: " +
ex.Message);
    }
}
```

The source code of the function that implements the above modeling algorithm is based on a miscalculation of one iteration of the CA.

```
public void _do_iteration()
{
```

```
    bits_counter ++;
    if (!backgroundWorker1.IsBusy)
    {
        label20.Text = bits_counter.ToString();
    }
    addBit_index = bits_counter % lenKA;
    if (addBit_index == 0)
    {
        addBit_index = lenKA;
    }
    addBit = KA[addBit_index];
    int prebit = addBit_index - 1;
    if (prebit <= 0)
    {
        prebit = lenKA;
    }
    if (!backgroundWorker1.IsBusy)
    {
        if (tb[prebit].Font.Bold)
        {
            tb[prebit].Font = new Font("Courier New", 10,
FontStyle.Regular);
        }
        tb[addBit_index].Font = new Font("Lucida Console", 12,
FontStyle.Bold);
        label13.Text = tb[addBit_index].Text;
    }
    int kol_actives = 0;
    for (int i = 0; i < 3; i++)
    {
        if (active_cell[i] != 0)
        {
            kol_actives++;
            int active_goto = _get_perehod(active_cell[i],
active_perehod[i]);
            Boolean active_goto_collision = false;
            for (int q = 0; q < 3; q++)
            {
                if (active_goto == active_cell[q])
                {
                    active_goto_collision = true;
                    break;
                }
            }
            if (active_goto_collision)
            {
                int[] s = _get_KA_sosedi(active_cell[i]);
                for (int k = 0; k < 6; k++)
                {
```

```
                    Boolean s_empty = true;
                    for (int q = 0; q < 3; q++)
                    {
                        if (s[k] == active_cell[q])
                        {
                            s_empty = false;
                        }
                    }
                    if (s_empty)
                    {
                        active_goto = s[k];
                        active_goto_collision = false;
                        break;
                    }
                }
            }
            if (!(active_goto_collision))
            {
                if (KA[active_cell[i]])
                {
                    tb[active_cell[i]].BackColor = color1;
                }
                else
                {
                    tb[active_cell[i]].BackColor = color0;
                }
                active_cell[i] = active_goto;
                if (!backgroundWorker1.IsBusy)
                {
                    tb[active_cell[i]].BackColor = colorA[i];
                }
                KA[active_cell[i]] = _get_bit(active_cell[i],
active_func[i]) ^ (checkBox1.Checked & addBit);
                if (!backgroundWorker1.IsBusy)
                {
                    tb[active_cell[i]].Text = Convert.
ToInt16(KA[active_cell[i]]).ToString();
                }
            }
        }
    }
    if (kol_actives < 2)
    {
        if (timer1.Enabled)
        {
            timer1.Stop();
            timer1.Enabled = false;
            button1.Text = "Start";
        }
```

```
        MessageBox.Show("To work, you must first activate two
cells: holding Ctrl, left-click on any cell of the CA and set
it the desired parameters ... "," We won't go there!)))");
        return;
    }
    if (kol_actives == 2)
    {
        int[] s = _get_KA_sosedi(active_cell[0]);
        for (int i = 0; i < 6; i++)
        {
            if (s[i] == active_cell[1])
            {
                active_cell[2] = s[(i + 3) % 6];
                if (!backgroundWorker1.IsBusy)
                {
                    tb[active_cell[2]].BackColor = colorA[2];
                }
                for (int q = 0; q < func_names.Length; q ++)
                {
                    if ((active_func[0] - 1 != q) && (active_
func[1] - 1 != q))
                    {
                        active_func[2] = q + 1;
                        break;
                    }
                }
                for (int q = 0; q < perehod_names.Length; q++)
                {
                    if ((active_perehod[0] - 1 != q) &&
(active_perehod[1] - 1 != q))
                    {
                        active_perehod[2] = q + 1;
                        break;
                    }
                }
                break;
            }
        }
    }
    int str_len_limit = 29;
    List<bool> abits = new List<bool>();
    StringBuilder shex = new StringBuilder(2);
    byte new_byte;
    int nb_int;
    shex.Append("  ");
    if (active_cell[0] != 0)
    {
        if (!backgroundWorker1.IsBusy)
        {
```

```
        label4.Text = tb[active_cell[0]].Text;
    }
abits.Add(KA[active_cell[0]]);
if (KA[active_cell[0]])
{
    active_bits_buf[0] = active_bits_buf[0] + "1";
}
else
{
    active_bits_buf[0] = active_bits_buf[0] + "0";
}
if (active_bits_buf[0].Length == 8)
{
    new_byte = Convert.ToByte(active_bits_buf[0], 2);
    active1_bytes.Add(new_byte);
    active_bits_buf[0] = string.Empty;
    if (!backgroundWorker1.IsBusy)
    {
        shex.Remove(0, 2);
        shex.AppendFormat("{0:x2}", active1_
bytes[active1_bytes.Count - 1]);
        label5.Text = label5.Text + '-' + shex.
ToString();
        if (label5.Text.Length > str_len_limit)
        {
            label5.Text = label5.Text.Remove(0, label5.
Text.Length - str_len_limit);
        }
    }
    if (((checkBox2.Checked) && (radioButton6.Checked))
|| (!radioButton6.Checked))
    {
        nb_int = Convert.ToInt32(new_byte);
        if ((active1_bytes.Count % 2) == 0)
        {
        graph_A12.FillRectangle(Brushes.Red,
Convert.ToInt32(active1_bytes[active1_bytes.Count - 2]), nb_
int, 1, 1);
            pictureBoxA12.Invalidate();
            if (!(preMal[1].IsEmpty))
            {
                graph_A12.FillRectangle(Brushes.Black,
preMal[1].X, preMal[1].Y, 1, 1);
                pictureBoxA12.Invalidate();
            }
            preMal[1].X = Convert.ToInt32(active1_
bytes[active1_bytes.Count - 2]);
            preMal[1].Y = nb_int;
        }
```

```
                A1_counter_byte[nb_int]++;
                if ((A1_counter_byte[nb_int] + 5) > max_
graph[1])
                {
                        max_graph[1] = max_graph[1] * 2 - max_
graph[1] / 2;
                        bm_A11 = new Bitmap(256, max_graph[1]);
                        pictureBoxA11.SizeMode =
PictureBoxSizeMode.StretchImage;
                        pictureBoxA11.Image = bm_A11;
                        graph_A11 = Graphics.
FromImage(pictureBoxA11.Image);
                        graph_A11.Clear(Color.White);
                        pictureBoxA11.Invalidate();
                        for (int i = 0; i < A1_counter_byte.Length;
i++)
                        {
                            graph_A11.DrawLine(Pens.Red, i, max_
graph[1], i, max_graph[1] - A1_counter_byte[i]);
                        }
                        pictureBoxA11.Invalidate();
                }
                else
                {
                        graph_A11.DrawLine(Pens.Red, nb_int, max_
graph[1], nb_int, max_graph[1] - A1_counter_byte[nb_int]);
                }
                pictureBoxA11.Invalidate();
            }
        }
    }
    if (active_cell[1] != 0)
    {
        if (!backgroundWorker1.IsBusy)
        {
            label7.Text = tb[active_cell[1]].Text;
        }
        abits.Add(KA[active_cell[1]]);
        if (KA[active_cell[1]])
        {
            active_bits_buf[1] = active_bits_buf[1] + "1";
        }
        else
        {
            active_bits_buf[1] = active_bits_buf[1] + "0";
        }
        if (active_bits_buf[1].Length == 8)
        {
            new_byte = Convert.ToByte(active_bits_buf[1], 2);
```

```
active2_bytes.Add(new_byte);
active_bits_buf[1] = string.Empty;
if (!backgroundWorker1.IsBusy)
{
    shex.Remove(0, 2);
    shex.AppendFormat("{0:x2}", active2_
bytes[active2_bytes.Count - 1]);
    label6.Text = label6.Text + '-' + shex.
ToString();
    if (label6.Text.Length > str_len_limit)
    {
        label6.Text = label6.Text.Remove(0, label6.
Text.Length - str_len_limit);
    }
}
if (((checkBox2.Checked) && (radioButton6.Checked))
|| (!radioButton6.Checked))
{
    nb_int = Convert.ToInt32(new_byte);
    if ((active2_bytes.Count % 2) == 0)
    {
        graph_A22.FillRectangle(Brushes.Red,
Convert.ToInt32(active2_bytes[active2_bytes.Count - 2]), nb_
int, 1, 1);
        pictureBoxA22.Invalidate();
        if (!(preMal[2].IsEmpty))
        {
            graph_A22.FillRectangle(Brushes.Black,
preMal[2].X, preMal[2].Y, 1, 1);
            pictureBoxA22.Invalidate();
        }
        preMal[2].X = Convert.ToInt32(active2_
bytes[active2_bytes.Count - 2]);
        preMal[2].Y = nb_int;
    }
    A2_counter_byte[nb_int]++;
    if ((A2_counter_byte[nb_int] + 5) > max_
graph[2])
    {
        max_graph[2] = max_graph[2] * 2 - max_
graph[2] / 2;
        bm_A21 = new Bitmap(256, max_graph[2]);
        pictureBoxA21.SizeMode =
PictureBoxSizeMode.StretchImage;
        pictureBoxA21.Image = bm_A21;
        graph_A21 = Graphics.
FromImage(pictureBoxA21.Image);
        graph_A21.Clear(Color.White);
        pictureBoxA21.Invalidate();
```

```
                for (int i = 0; i < A2_counter_byte.Length;
i++)
                    {
                        graph_A21.DrawLine(Pens.OrangeRed, i,
max_graph[2], i, max_graph[2] - A2_counter_byte[i]);
                    }
                    pictureBoxA21.Invalidate();
                }
                else
                {
                    graph_A21.DrawLine(Pens.OrangeRed, nb_int,
max_graph[2], nb_int, max_graph[2] - A2_counter_byte[nb_int]);
                }
                pictureBoxA21.Invalidate();
            }
        }
    }
    if (active_cell[2] != 0)
    {
        if (!backgroundWorker1.IsBusy)
        {
            label10.Text = tb[active_cell[2]].Text;
        }
        abits.Add(KA[active_cell[2]]);
        if (KA[active_cell[2]])
        {
            active_bits_buf[2] = active_bits_buf[2] + "1";
        }

        else
        {
            active_bits_buf[2] = active_bits_buf[2] + "0";
        }
        if (active_bits_buf[2].Length == 8)
        {
            new_byte = Convert.ToByte(active_bits_buf[2], 2);
            active3_bytes.Add(new_byte);
            active_bits_buf[2] = string.Empty;
            if (!backgroundWorker1.IsBusy)
            {
                shex.Remove(0, 2);
                shex.AppendFormat("{0:x2}", active3_
bytes[active3_bytes.Count - 1]);
                label9.Text = label9.Text + '-' + shex.
ToString();
                if (label9.Text.Length > str_len_limit)
                {
                    label9.Text = label9.Text.Remove(0, label9.
Text.Length - str_len_limit);
```

```
                }
            }
        if (((checkBox2.Checked) && (radioButton6.Checked))
|| (!radioButton6.Checked))
            {
                nb_int = Convert.ToInt32(new_byte);
                if ((active3_bytes.Count % 2) == 0)
                {
                    graph_A32.FillRectangle(Brushes.Red,
Convert.ToInt32(active3_bytes[active3_bytes.Count - 2]), nb_
int, 1, 1);
                    pictureBoxA32.Invalidate();
                    if (!(preMal[3].IsEmpty))
                    {
                        graph_A32.FillRectangle(Brushes.Black,
preMal[3].X, preMal[3].Y, 1, 1);
                        pictureBoxA32.Invalidate();
                    }
                    preMal[3].X = Convert.ToInt32(active3_
bytes[active3_bytes.Count - 2]);
                    preMal[3].Y = nb_int;
                }
                A3_counter_byte[nb_int]++;
                if ((A3_counter_byte[nb_int] + 5) > max_
graph[3])
                {
                    max_graph[3] = max_graph[3] * 2 - max_
graph[3] / 2;
                    bm_A31 = new Bitmap(256, max_graph[3]);
                    pictureBoxA31.SizeMode =
PictureBoxSizeMode.StretchImage;
                    pictureBoxA31.Image = bm_A31;
                    graph_A31 = Graphics.
FromImage(pictureBoxA31.Image);
                    graph_A31.Clear(Color.White);
                    pictureBoxA31.Invalidate();
                    for (int i = 0; i < A3_counter_byte.Length;
i++)
                    {
                        graph_A31.DrawLine(Pens.LightCoral, i,
max_graph[3], i, max_graph[3] - A3_counter_byte[i]);
                    }
                    pictureBoxA31.Invalidate();
                }
                else
                {
                    graph_A31.DrawLine(Pens.LightCoral, nb_int,
max_graph[3], nb_int, max_graph[3] - A3_counter_byte[nb_int]);
                }
```

```
                pictureBoxA31.Invalidate();
        }
    }
}
Boolean b = false;
if (abits.Count > 1)
{
    b = abits[0];
    for (int i = 1; i < abits.Count; i++)
    {
        b = b ^ abits[i];
    }
}
if (!backgroundWorker1.IsBusy)
{
    label16.Text = Convert.ToInt16(b).ToString();
}
RND_bits_buf = RND_bits_buf + Convert.ToInt16(b);
if (RND_bits_buf.Length == 8)
{
    new_byte = Convert.ToByte(RND_bits_buf, 2);
    RND_bytes.Add(new_byte);
    RND_bits_buf = string.Empty;
    if (!backgroundWorker1.IsBusy)
    {
        label23.Text = RND_bytes.Count.ToString();
        str_len_limit = 26;
        shex.Remove(0, 2);
        shex.AppendFormat("{0:x2}", new_byte);
        label15.Text = label15.Text + '-' + shex.
ToString();
        if (label15.Text.Length > str_len_limit)
        {
            label15.Text = label15.Text.Remove(0, label15.
Text.Length - str_len_limit);
        }
    }
    if (((checkBox2.Checked) && (radioButton6.Checked)) ||
(!radioButton6.Checked))
    {
        nb_int = Convert.ToInt32(new_byte);
        if ((RND_bytes.Count % 2) == 0)
        {
            graph_RND2.FillRectangle(Brushes.Red, Convert.
ToInt32(RND_bytes[RND_bytes.Count - 2]), nb_int, 1, 1);
            pictureBox2.Invalidate();
            if (!(preMal[0].IsEmpty))
            {
```

```
                         graph_RND2.FillRectangle(Brushes.Black,
preMal[0].X, preMal[0].Y, 1, 1);
                    pictureBox2.Invalidate();
                }
                preMal[0].X = Convert.ToInt32(RND_bytes[RND_
bytes.Count - 2]);
                preMal[0].Y = nb_int;
            }
            RND_counter_byte[nb_int]++;
            if ((RND_counter_byte[nb_int] + 5) > max_graph[0])
            {
                max_graph[0] = max_graph[0] * 2 - max_graph[0]
/ 2;
                bm_RND1 = new Bitmap(256, max_graph[0]);
                pictureBox1.SizeMode = PictureBoxSizeMode.
StretchImage;
                pictureBox1.Image = bm_RND1;
                graph_RND1 = Graphics.FromImage(pictureBox1.
Image);
                graph_RND1.Clear(Color.White);
                pictureBox1.Invalidate();
                for (int i = 0; i < RND_counter_byte.Length;
i++)
                {
                    graph_RND1.DrawLine(Pens.Green, i, max_
graph[0], i, max_graph[0] - RND_counter_byte[i]);
                }
                pictureBox1.Invalidate();
            }
            else
            {
                graph_RND1.DrawLine(Pens.Green, nb_int, max_
graph[0], nb_int, max_graph[0] - RND_counter_byte[nb_int]);
            }
            pictureBox1.Invalidate();
        }
    }
}
```

The layout of the controls on the main form reproduces the listing of the file Form1.Designer.cs:

```
namespace HexRND3
{
    partial class Form1
    {
        /// <summary>
        /// Required designer variable.
```

```csharp
        /// </summary>
        private System.ComponentModel.IContainer components =
null;
        /// <summary>
        /// Clean up any resources being used.
        /// </summary>
        /// <param name="disposing">true if managed resources
should be disposed; otherwise, false.</param>
        protected override void Dispose(bool disposing)
        {
            if (disposing && (components != null))
            {
                components.Dispose();
            }
            base.Dispose(disposing);
        }
        #region Windows Form Designer generated code
        /// <summary>
        /// Required method for Designer support - do not
modify
        /// the contents of this method with the code editor.
        /// </summary>
        private void InitializeComponent()
        {
            this.components = new System.ComponentModel.
Container();
            System.ComponentModel.ComponentResourceManager
resources = new System.ComponentModel.ComponentResourceManager(
typeof(Form1));
            this.button1 = new System.Windows.Forms.Button();
            this.groupBox4 = new System.Windows.Forms.
GroupBox();
            this.label2 = new System.Windows.Forms.Label();
            this.groupBox1 = new System.Windows.Forms.
GroupBox();
            this.radioButton3 = new System.Windows.Forms.
RadioButton();
            this.radioButton2 = new System.Windows.Forms.
RadioButton();
            this.radioButton1 = new System.Windows.Forms.
RadioButton();
            this.button2 = new System.Windows.Forms.Button();
            this.label21 = new System.Windows.Forms.Label();
            this.label19 = new System.Windows.Forms.Label();
            this.comboBox5 = new System.Windows.Forms.
ComboBox();
            this.comboBox4 = new System.Windows.Forms.
ComboBox();
```

```
            this.comboBox3 = new System.Windows.Forms.
ComboBox();
            this.panel1 = new System.Windows.Forms.Panel();
            this.label1 = new System.Windows.Forms.Label();
            this.backgroundWorker1 = new System.ComponentModel.
BackgroundWorker();
            this.toolTip1 = new System.Windows.Forms.
ToolTip(this.components);
            this.textBox15 = new System.Windows.Forms.
TextBox();
            this.radioButton6 = new System.Windows.Forms.
RadioButton();
            this.textBox13 = new System.Windows.Forms.
TextBox();
            this.radioButton4 = new System.Windows.Forms.
RadioButton();
            this.radioButton5 = new System.Windows.Forms.
RadioButton();
            this.groupBox2 = new System.Windows.Forms.
GroupBox();
            this.button4 = new System.Windows.Forms.Button();
            this.groupBox5 = new System.Windows.Forms.
GroupBox();
            this.checkBox2 = new System.Windows.Forms.
CheckBox();
            this.checkBox1 = new System.Windows.Forms.
CheckBox();
            this.label15 = new System.Windows.Forms.Label();
            this.label16 = new System.Windows.Forms.Label();
            this.label17 = new System.Windows.Forms.Label();
            this.label13 = new System.Windows.Forms.Label();
            this.label14 = new System.Windows.Forms.Label();
            this.label9 = new System.Windows.Forms.Label();
            this.label10 = new System.Windows.Forms.Label();
            this.label11 = new System.Windows.Forms.Label();
            this.label6 = new System.Windows.Forms.Label();
            this.label7 = new System.Windows.Forms.Label();
            this.label8 = new System.Windows.Forms.Label();
            this.label5 = new System.Windows.Forms.Label();
            this.label4 = new System.Windows.Forms.Label();
            this.label3 = new System.Windows.Forms.Label();
            this.pictureBox2 = new System.Windows.Forms.
PictureBox();
            this.label23 = new System.Windows.Forms.Label();
            this.label22 = new System.Windows.Forms.Label();
            this.label20 = new System.Windows.Forms.Label();
            this.label18 = new System.Windows.Forms.Label();
            this.timer1 = new System.Windows.Forms.Timer(this.
components);
```

```
            this.menuStrip1 = new System.Windows.Forms.
MenuStrip();
            this.fileToolStripMenuItem = new System.Windows.
Forms.ToolStripMenuItem();
            this.saveConditionCameExternalFileToolStripMenuItem
= new System.Windows.Forms.ToolStripMenuItem();
            this.
DownloadConditionFromanExternalFileToolStripMenuItem = new
System.Windows.Forms.ToolStripMenuItem();
            this.toolStripSeparator1 = new System.Windows.
Forms.ToolStripSeparator();
            this.ExitfromProgramToolStripMenuItem = new System.
Windows.Forms.ToolStripMenuItem();
            this.referenceToolStripMenuItem = new System.
Windows.Forms.ToolStripMenuItem();
            this.aboutTheProgramToolStripMenuItem = new System.
Windows.Forms.ToolStripMenuItem();
            this.colorDialog1 = new System.Windows.Forms.
ColorDialog();
            this.saveFileDialog1 = new System.Windows.Forms.
SaveFileDialog();
            this.openFileDialog1 = new System.Windows.Forms.
OpenFileDialog();
            this.button3 = new System.Windows.Forms.Button();
            this.tabPage8 = new System.Windows.Forms.TabPage();
            this.pictureBox6 = new System.Windows.Forms.
PictureBox();
            this.pictureBox7 = new System.Windows.Forms.
PictureBox();
            this.label41 = new System.Windows.Forms.Label();
            this.label42 = new System.Windows.Forms.Label();
            this.label43 = new System.Windows.Forms.Label();
            this.label44 = new System.Windows.Forms.Label();
            this.label45 = new System.Windows.Forms.Label();
            this.label46 = new System.Windows.Forms.Label();
            this.label47 = new System.Windows.Forms.Label();
            this.tabPage9 = new System.Windows.Forms.TabPage();
            this.pictureBox8 = new System.Windows.Forms.
PictureBox();
            this.pictureBox9 = new System.Windows.Forms.
PictureBox();
            this.label48 = new System.Windows.Forms.Label();
            this.label49 = new System.Windows.Forms.Label();
            this.label50 = new System.Windows.Forms.Label();
            this.label51 = new System.Windows.Forms.Label();
            this.label52 = new System.Windows.Forms.Label();
            this.label53 = new System.Windows.Forms.Label();
            this.label54 = new System.Windows.Forms.Label();
            this.tabPage2 = new System.Windows.Forms.TabPage();
```

```
            this.tabControl2 = new System.Windows.Forms.
TabControl();
            this.tabPage11 = new System.Windows.Forms.
TabPage();
            this.pictureBox1 = new System.Windows.Forms.
PictureBox();
            this.label62 = new System.Windows.Forms.Label();
            this.label63 = new System.Windows.Forms.Label();
            this.label64 = new System.Windows.Forms.Label();
            this.label65 = new System.Windows.Forms.Label();
            this.label66 = new System.Windows.Forms.Label();
            this.label67 = new System.Windows.Forms.Label();
            this.label68 = new System.Windows.Forms.Label();
            this.tabPage6 = new System.Windows.Forms.TabPage();
            this.pictureBoxA11 = new System.Windows.Forms.
PictureBox();
            this.pictureBoxA12 = new System.Windows.Forms.
PictureBox();
            this.label30 = new System.Windows.Forms.Label();
            this.label27 = new System.Windows.Forms.Label();
            this.label31 = new System.Windows.Forms.Label();
            this.label29 = new System.Windows.Forms.Label();
            this.label32 = new System.Windows.Forms.Label();
            this.label28 = new System.Windows.Forms.Label();
            this.label33 = new System.Windows.Forms.Label();
            this.tabPage7 = new System.Windows.Forms.TabPage();
            this.pictureBoxA22 = new System.Windows.Forms.
PictureBox();
            this.pictureBoxA21 = new System.Windows.Forms.
PictureBox();
            this.label34 = new System.Windows.Forms.Label();
            this.label35 = new System.Windows.Forms.Label();
            this.label36 = new System.Windows.Forms.Label();
            this.label37 = new System.Windows.Forms.Label();
            this.label38 = new System.Windows.Forms.Label();
            this.label39 = new System.Windows.Forms.Label();
            this.label40 = new System.Windows.Forms.Label();
            this.tabPage10 = new System.Windows.Forms.
TabPage();
            this.pictureBoxA32 = new System.Windows.Forms.
PictureBox();
            this.pictureBoxA31 = new System.Windows.Forms.
PictureBox();
            this.label55 = new System.Windows.Forms.Label();
            this.label56 = new System.Windows.Forms.Label();
            this.label57 = new System.Windows.Forms.Label();
            this.label58 = new System.Windows.Forms.Label();
            this.label59 = new System.Windows.Forms.Label();
            this.label60 = new System.Windows.Forms.Label();
```

```
            this.label61 = new System.Windows.Forms.Label();
            this.tabPage1 = new System.Windows.Forms.TabPage();
            this.groupBox6 = new System.Windows.Forms.
GroupBox();
            this.tabControl1 = new System.Windows.Forms.
TabControl();
            this.progressBar1 = new System.Windows.Forms.
ProgressBar();
            this.timer2 = new System.Windows.Forms.Timer(this.
components);
            this.panel2 = new System.Windows.Forms.Panel();
            this.label75 = new System.Windows.Forms.Label();
            this.label73 = new System.Windows.Forms.Label();
            this.label72 = new System.Windows.Forms.Label();
            this.label74 = new System.Windows.Forms.Label();
            this.label69 = new System.Windows.Forms.Label();
            this.label70 = new System.Windows.Forms.Label();
            this.label71 = new System.Windows.Forms.Label();
```
        this.label26 = new System.Windows.Forms.Label();
```
            this.label25 = new System.Windows.Forms.Label();
            this.label24 = new System.Windows.Forms.Label();
            this.label12 = new System.Windows.Forms.Label();
            this.groupBox4.SuspendLayout();
            this.groupBox1.SuspendLayout();
            this.groupBox2.SuspendLayout();
            this.groupBox5.SuspendLayout();
            ((System.ComponentModel.ISupportInitialize)(this.
pictureBox2)).BeginInit();
            this.menuStrip1.SuspendLayout();
            this.tabPage8.SuspendLayout();
            ((System.ComponentModel.ISupportInitialize)(this.
pictureBox6)).BeginInit();
            ((System.ComponentModel.ISupportInitialize)(this.
pictureBox7)).BeginInit();
            this.tabPage9.SuspendLayout();
            ((System.ComponentModel.ISupportInitialize)(this.
pictureBox8)).BeginInit();
            ((System.ComponentModel.ISupportInitialize)(this.
pictureBox9)).BeginInit();
            this.tabPage2.SuspendLayout();
            this.tabControl2.SuspendLayout();
            this.tabPage11.SuspendLayout();
            ((System.ComponentModel.ISupportInitialize)(this.
pictureBox1)).BeginInit();
            this.tabPage6.SuspendLayout();
            ((System.ComponentModel.ISupportInitialize)(this.
pictureBoxA11)).BeginInit();
            ((System.ComponentModel.ISupportInitialize)(this.
pictureBoxA12)).BeginInit();
```

```
            this.tabPage7.SuspendLayout();
            ((System.ComponentModel.ISupportInitialize)(this.
pictureBoxA22)).BeginInit();
            ((System.ComponentModel.ISupportInitialize)(this.
pictureBoxA21)).BeginInit();
            this.tabPage10.SuspendLayout();
            ((System.ComponentModel.ISupportInitialize)(this.
pictureBoxA32)).BeginInit();
            ((System.ComponentModel.ISupportInitialize)(this.
pictureBoxA31)).BeginInit();
            this.tabPage1.SuspendLayout();
            this.groupBox6.SuspendLayout();
            this.tabControl1.SuspendLayout();
            this.panel2.SuspendLayout();
            this.SuspendLayout();
            //
            // button1
            //
            this.button1.Location = new System.Drawing.
Point(180, 131);
            this.button1.Name = "button1";
            this.button1.Size = new System.Drawing.Size(94,
25);
            this.button1.TabIndex = 0;
            this.button1.Text = "Start";
            this.button1.UseVisualStyleBackColor = true;
            this.button1.Click += new System.EventHandler(this.
button1_Click);
            //
            // groupBox4
            //
            this.groupBox4.Controls.Add(this.label2);
            this.groupBox4.Controls.Add(this.groupBox1);
            this.groupBox4.Controls.Add(this.button2);
            this.groupBox4.Controls.Add(this.label21);
            this.groupBox4.Controls.Add(this.label19);
            this.groupBox4.Controls.Add(this.comboBox5);
            this.groupBox4.Controls.Add(this.comboBox4);
            this.groupBox4.Controls.Add(this.comboBox3);
            this.groupBox4.Location = new System.Drawing.
Point(8, 18);
            this.groupBox4.Name = "groupBox4";
            this.groupBox4.Size = new System.Drawing.Size(300,
89);
            this.groupBox4.TabIndex = 27;
            this.groupBox4.TabStop = false;
            this.groupBox4.Text = "CA Constructor ";
            //
            // label2
```

```
            //
            this.label2.AutoSize = true;
            this.label2.Font = new System.Drawing.
Font("Microsoft Sans Serif", 8.25F, System.Drawing.FontStyle.
Regular, System.Drawing.GraphicsUnit.Point, ((byte)(204)));
            this.label2.Location = new System.Drawing.Point(6,
40);
            this.label2.Name = "label2";
            this.label2.Size = new System.Drawing.Size(32, 13);
            this.label2.TabIndex = 4;
            this.label2.Text = "by Y:";
            this.label2.TextAlign = System.Drawing.
ContentAlignment.MiddleRight;
            //
            // groupBox1
            //
            this.groupBox1.Controls.Add(this.radioButton3);
            this.groupBox1.Controls.Add(this.radioButton2);
            this.groupBox1.Controls.Add(this.radioButton1);
            this.groupBox1.Location = new System.Drawing.
Point(87, 10);
            this.groupBox1.Name = "groupBox1";
            this.groupBox1.Size = new System.Drawing.Size(88,
69);
            this.groupBox1.TabIndex = 3;
            this.groupBox1.TabStop = false;
            this.groupBox1.Text = "filling out ";
            //
            // radioButton3
            //
            this.radioButton3.AutoSize = true;
            this.radioButton3.Checked = true;
            this.radioButton3.Location = new System.Drawing.
Point(6, 18);
            this.radioButton3.Name = "radioButton3";
            this.radioButton3.Size = new System.Drawing.
Size(71, 17);
            this.radioButton3.TabIndex = 2;
            this.radioButton3.TabStop = true;
            this.radioButton3.Text = "randomly";
            this.radioButton3.UseVisualStyleBackColor = true;
            this.radioButton3.Click += new System.
EventHandler(this.radioButton3_Click);
            this.radioButton3.CheckedChanged += new System.
EventHandler(this.radioButton3_CheckedChanged);
            //
            // radioButton2
            //
            this.radioButton2.AutoSize = true;
```

```
            this.radioButton2.Location = new System.Drawing.
Point(6, 48);
            this.radioButton2.Name = "radioButton2";
            this.radioButton2.Size = new System.Drawing.
Size(41, 17);
            this.radioButton2.TabIndex = 1;
            this.radioButton2.Text = "\"1\"";
            this.radioButton2.UseVisualStyleBackColor = true;
            this.radioButton2.CheckedChanged += new System.
EventHandler(this.radioButton2_CheckedChanged);
            //
            // radioButton1
            //
            this.radioButton1.AutoSize = true;
            this.radioButton1.Location = new System.Drawing.
Point(6, 33);
            this.radioButton1.Name = "radioButton1";
            this.radioButton1.Size = new System.Drawing.
Size(41, 17);
            this.radioButton1.TabIndex = 0;
            this.radioButton1.Text = "\"0\"";
            this.radioButton1.UseVisualStyleBackColor = true;
            this.radioButton1.CheckedChanged += new System.
EventHandler(this.radioButton1_CheckedChanged);
            //
            // button2
            //
            this.button2.Location = new System.Drawing.
Point(181, 30);
            this.button2.Name = "button2";
            this.button2.Size = new System.Drawing.Size(102,
43);
            this.button2.TabIndex = 2;
            this.button2.Text = "CA initialization";
            this.button2.UseVisualStyleBackColor = true;
            this.button2.Click += new System.EventHandler(this.
button2_Click);
            //
            // label21
            //
            this.label21.BackColor = System.Drawing.
SystemColors.Control;
            this.label21.Font = new System.Drawing.
Font("Microsoft Sans Serif", 6.75F, System.Drawing.FontStyle.
Regular, System.Drawing.GraphicsUnit.Point, ((byte)(204)));
            this.label21.ForeColor = System.Drawing.
SystemColors.ControlText;
            this.label21.Location = new System.Drawing.Point(4,
55);
```

```
            this.label21.Name = "label21";
            this.label21.Size = new System.Drawing.Size(37,
24);
            this.label21.TabIndex = 1;
            this.label21.Text = "cell size ";
            this.label21.TextAlign = System.Drawing.
ContentAlignment.MiddleRight;
            //
            // label19
            //
            this.label19.AutoSize = true;
            this.label19.Font = new System.Drawing.
Font("Microsoft Sans Serif", 8.25F, System.Drawing.FontStyle.
Regular, System.Drawing.GraphicsUnit.Point, ((byte)(204)));
            this.label19.Location = new System.Drawing.Point(6, 21);
            this.label19.Name = "label19";
            this.label19.Size = new System.Drawing.Size(32,
13);
            this.label19.TabIndex = 1;
            this.label19.Text = "by X:";
            this.label19.TextAlign = System.Drawing.
ContentAlignment.MiddleRight;
            //
            // comboBox5
            //
            this.comboBox5.DropDownStyle = System.Windows.
Forms.ComboBoxStyle.DropDownList;
            this.comboBox5.FormattingEnabled = true;
            this.comboBox5.Items.AddRange(new object[] {
            "8",
            "9",
            "10",
            "11",
            "12",
            "13",
            "14",
            "15",
            "16",
            "17",
            "18",
            "19",
            "20",
            "21",
            "22",
            "23",
            "24",
            "25",
            "26",
            "27",
```

```
            "28",
            "29",
            "30",
            "31",
            "32",
            "33",
            "34",
            "35",
            "36",
            "37",
            "38",
            "39",
            "40"});
            this.comboBox5.Location = new System.Drawing.
Point(45, 58);
            this.comboBox5.Name = "comboBox5";
            this.comboBox5.Size = new System.Drawing.Size(36,
21);
            this.comboBox5.TabIndex = 0;
            //
            // comboBox4
            //
            this.comboBox4.DropDownStyle = System.Windows.
Forms.ComboBoxStyle.DropDownList;
            this.comboBox4.FormattingEnabled = true;
            this.comboBox4.Items.AddRange(new object[] {
            "4",
            "6",
            "8",
            "10",
            "12",
            "14",
            "16",
            "18",
            "20",
            "22",
            "24",
            "26",
            "28",
            "30",
            "32",
            "34",
            "36",
            "38",
            "40"});
            this.comboBox4.Location = new System.Drawing.
Point(45, 37);
            this.comboBox4.Name = "comboBox4";
```

```
            this.comboBox4.Size = new System.Drawing.Size(36,
21);
            this.comboBox4.TabIndex = 0;
            //
            // comboBox3
            //
            this.comboBox3.DropDownStyle = System.Windows.
Forms.ComboBoxStyle.DropDownList;
            this.comboBox3.FormattingEnabled = true;
            this.comboBox3.Items.AddRange(new object[] {
            "3",
            "4",
            "5",
            "6",
            "7",
            "8",
            "9",
            "10",
            "11",
            "12",
            "13",
            "14",
            "15",
            "16",
            "17",
            "18",
            "19",
            "20",
            "21",
            "22",
            "23",
            "24",
            "25",
            "26",
            "27",
            "28",
            "29",
            "30",
            "31",
            "32",
            "33",
            "34",
            "35",
            "36",
            "37",
            "38",
            "39",
            "40"});
```

```
            this.comboBox3.Location = new System.Drawing.
Point(45, 16);
            this.comboBox3.Name = "comboBox3";
            this.comboBox3.Size = new System.Drawing.Size(36,
21);
            this.comboBox3.TabIndex = 0;
            //
            // panel1
            //
            this.panel1.AutoScroll = true;
            this.panel1.AutoSize = true;
            this.panel1.BackColor = System.Drawing.Color.
SlateGray;
            this.panel1.BorderStyle = System.Windows.Forms.
BorderStyle.FixedSingle;
            this.panel1.Cursor = System.Windows.Forms.Cursors.
Default;
            this.panel1.Location = new System.Drawing.
Point(332, 37);
            this.panel1.Margin = new System.Windows.Forms.
Padding(1, 1, 3, 3);
            this.panel1.MinimumSize = new System.Drawing.
Size(80, 80);
            this.panel1.Name = "panel1";
            this.panel1.Size = new System.Drawing.Size(285,
107);
            this.panel1.TabIndex = 28;
            //
            // label1
            //
            this.label1.BackColor = System.Drawing.
SystemColors.Control;
            this.label1.Font = new System.Drawing.
Font("Microsoft Sans Serif", 9.75F, System.Drawing.FontStyle.
Bold, System.Drawing.GraphicsUnit.Point, ((byte)(204)));
            this.label1.Location = new System.Drawing.
Point(334, 147);
            this.label1.Name = "label1";
            this.label1.Size = new System.Drawing.Size(237,
63);
            this.label1.TabIndex = 0;
            this.label1.Text = "... wait, initialization of a
new instance of the CA will be performed...";
            this.label1.TextAlign = System.Drawing.
ContentAlignment.MiddleCenter;
            //
            // backgroundWorker1
            //
```

```
            this.backgroundWorker1.WorkerReportsProgress =
true;
            this.backgroundWorker1.WorkerSupportsCancellation =
true;
            this.backgroundWorker1.DoWork += new System.
ComponentModel.DoWorkEventHandler(this.backgroundWorker1_
DoWork);
            this.backgroundWorker1.ProgressChanged += new
System.ComponentModel.ProgressChangedEventHandler(this.
backgroundWorker1_ProgressChanged);
            //
            // textBox15
            //
            this.textBox15.Enabled = false;
            this.textBox15.Location = new System.Drawing.
Point(144, 47);
            this.textBox15.Margin = new System.Windows.Forms.
Padding(1);
            this.textBox15.Name = "textBox15";
            this.textBox15.Size = new System.Drawing.Size(61,
20);
            this.textBox15.TabIndex = 18;
            this.textBox15.Text = "1048576";
            this.textBox15.TextAlign = System.Windows.Forms.
HorizontalAlignment.Center;
            this.toolTip1.SetToolTip(this.textBox15, "the
number of bytes generated in the statistics collection mode
(maximum for one session 67108864)");
            //
            // radioButton6
            //
            this.radioButton6.AutoSize = true;
            this.radioButton6.Location = new System.Drawing.
Point(6, 47);
            this.radioButton6.Name = "radioButton6";
            this.radioButton6.Size = new System.Drawing.
Size(141, 17);
            this.radioButton6.TabIndex = 17;
            this.radioButton6.TabStop = true;
            this.radioButton6.Text = "statistics collection
(bytes)";
            this.toolTip1.SetToolTip(this.radioButton6,
"the generation process is performed automatically with the
recording of all generated bytes in the corresponding files,
there is no graphic accompaniment");
            this.radioButton6.UseVisualStyleBackColor = true;
            this.radioButton6.CheckedChanged += new System.
EventHandler(this.radioButton6_CheckedChanged);
            //
```

```
// textBox13
//
this.textBox13.Location = new System.Drawing.
Point(159, 11);
this.textBox13.Name = "textBox13";
this.textBox13.Size = new System.Drawing.Size(45,
20);
this.textBox13.TabIndex = 16;
this.textBox13.Text = "20";
this.textBox13.TextAlign = System.Windows.Forms.
HorizontalAlignment.Center;
this.toolTip1.SetToolTip(this.textBox13, "interval
between iterations for automatic mode");
this.textBox13.TextChanged += new System.
EventHandler(this.textBox13_TextChanged);
//
// radioButton4
//
this.radioButton4.AutoSize = true;
this.radioButton4.Checked = true;
this.radioButton4.Location = new System.Drawing.
Point(6, 12);
this.radioButton4.Name = "radioButton4";
this.radioButton4.Size = new System.Drawing.
Size(147, 17);
this.radioButton4.TabIndex = 10;
this.radioButton4.TabStop = true;
this.radioButton4.Text = "auto (with an interval,
ms)";
this.toolTip1.SetToolTip(this.radioButton4, "the
generation process is performed automatically at the designated
interval, accompanied by a graphical interpretation of the
calculations");
this.radioButton4.UseVisualStyleBackColor = true;
//
// radioButton5
//
this.radioButton5.AutoSize = true;
this.radioButton5.Location = new System.Drawing.
Point(6, 30);
this.radioButton5.Name = "radioButton5";
this.radioButton5.Size = new System.Drawing.
Size(74, 17);
this.radioButton5.TabIndex = 10;
this.radioButton5.Text = "step by step";
this.toolTip1.SetToolTip(this.radioButton5, "the
generation of the next bit is performed at the touch of a
button, accompanied by a graphical interpretation of the
calculations");
```

```
                this.radioButton5.UseVisualStyleBackColor = true;
                //
                // groupBox2
                //
                this.groupBox2.Controls.Add(this.button4);
                this.groupBox2.Controls.Add(this.groupBox5);
                this.groupBox2.Controls.Add(this.checkBox1);
                this.groupBox2.Controls.Add(this.button1);
                this.groupBox2.Location = new System.Drawing.
Point(8, 122);
                this.groupBox2.Name = "groupBox2";
                this.groupBox2.Size = new System.Drawing.Size(300,
168);
                this.groupBox2.TabIndex = 29;
                this.groupBox2.TabStop = false;
                this.groupBox2.Text = "Mode of operation";
                //
                // button4
                //
                this.button4.Location = new System.Drawing.Point(6,
131);
                this.button4.Name = "button4";
                this.button4.Size = new System.Drawing.Size(136,
25);
                this.button4.TabIndex = 31;
                this.button4.Text = "Reset Results";
                this.button4.UseVisualStyleBackColor = true;
                this.button4.Click += new System.EventHandler(this.
button4_Click);
                //
                // groupBox5
                //
                this.groupBox5.Controls.Add(this.checkBox2);
                this.groupBox5.Controls.Add(this.textBox15);
                this.groupBox5.Controls.Add(this.radioButton6);
                this.groupBox5.Controls.Add(this.textBox13);
                this.groupBox5.Controls.Add(this.radioButton4);
                this.groupBox5.Controls.Add(this.radioButton5);
                this.groupBox5.Location = new System.Drawing.
Point(6, 42);
                this.groupBox5.Name = "groupBox5";
                this.groupBox5.Size = new System.Drawing.Size(286,
75);
                this.groupBox5.TabIndex = 30;
                this.groupBox5.TabStop = false;
                this.groupBox5.Text = "start";
                //
                // checkBox2
                //
```

```
            this.checkBox2.AutoSize = true;
            this.checkBox2.Checked = true;
            this.checkBox2.CheckState = System.Windows.Forms.
CheckState.Checked;
            this.checkBox2.Enabled = false;
            this.checkBox2.Location = new System.Drawing.
Point(214, 48);
            this.checkBox2.Name = "checkBox2";
            this.checkBox2.Size = new System.Drawing.Size(69,
17);
            this.checkBox2.TabIndex = 19;
            this.checkBox2.Text = "graphics";
            this.checkBox2.UseVisualStyleBackColor = true;
            //
            // checkBox1
            //
            this.checkBox1.AutoSize = true;
            this.checkBox1.Checked = true;
            this.checkBox1.CheckState = System.Windows.Forms.
CheckState.Checked;
            this.checkBox1.Location = new System.Drawing.
Point(13, 19);
            this.checkBox1.Name = "checkBox1";
            this.checkBox1.Size = new System.Drawing.Size(236,
17);
            this.checkBox1.TabIndex = 30;
            this.checkBox1.Text = "take into account add. bit
like XOR in every AC";
            this.checkBox1.UseVisualStyleBackColor = true;
            //
            // label15
            //
            this.label15.BorderStyle = System.Windows.Forms.
BorderStyle.FixedSingle;
            this.label15.Font = new System.Drawing.Font("Lucida
Console", 8.25F, System.Drawing.FontStyle.Bold, System.Drawing.
GraphicsUnit.Point, ((byte)(204)));
            this.label15.Location = new System.Drawing.
Point(49, 164);
            this.label15.Name = "label15";
            this.label15.Size = new System.Drawing.Size(225,
18);
            this.label15.TabIndex = 46;
            this.label15.TextAlign = System.Drawing.
ContentAlignment.MiddleCenter;
            //
            // label16
            //
```

```
                this.label16.BackColor = System.Drawing.
SystemColors.Control;
                this.label16.Font = new System.Drawing.
Font("Microsoft Sans Serif", 8.25F, System.Drawing.FontStyle.
Bold, System.Drawing.GraphicsUnit.Point, ((byte)(204)));
                this.label16.ForeColor = System.Drawing.Color.
Green;
                this.label16.Location = new System.Drawing.
Point(278, 165);
                this.label16.Name = "label16";
                this.label16.Size = new System.Drawing.Size(14,
19);
                this.label16.TabIndex = 45;
                this.label16.Text = "x";
                this.label16.TextAlign = System.Drawing.
ContentAlignment.MiddleCenter;
                //
                // label17
                //
                this.label17.AutoSize = true;
                this.label17.Font = new System.Drawing.
Font("Microsoft Sans Serif", 8.25F, System.Drawing.FontStyle.
Bold, System.Drawing.GraphicsUnit.Point, ((byte)(204)));
                this.label17.Location = new System.Drawing.
Point(12, 166);
                this.label17.Name = "label17";
                this.label17.Size = new System.Drawing.Size(37,
13);
                this.label17.TabIndex = 44;
                this.label17.Text = "XOR:";
                //
                // label13
                //
                this.label13.Font = new System.Drawing.
Font("Microsoft Sans Serif", 8.25F, System.Drawing.FontStyle.
Bold, System.Drawing.GraphicsUnit.Point, ((byte)(204)));
                this.label13.ForeColor = System.Drawing.Color.
SaddleBrown;
                this.label13.Location = new System.Drawing.
Point(278, 24);
                this.label13.Name = "label13";
                this.label13.Size = new System.Drawing.Size(14,
19);
                this.label13.TabIndex = 42;
                this.label13.Text = "x";
                this.label13.TextAlign = System.Drawing.
ContentAlignment.MiddleCenter;
                //
                // label14
```

```
            //
            this.label14.AutoSize = true;
            this.label14.Location = new System.Drawing.
Point(142, 28);
            this.label14.Name = "label14";
            this.label14.Size = new System.Drawing.Size(115,
13);
            this.label14.TabIndex = 41;
            this.label14.Text = "additional bit:";
            //
            // label9
            //
            this.label9.BorderStyle = System.Windows.Forms.
BorderStyle.FixedSingle;
            this.label9.Font = new System.Drawing.Font("Lucida
Console", 8.25F, System.Drawing.FontStyle.Regular, System.
Drawing.GraphicsUnit.Point, ((byte)(204)));
            this.label9.Location = new System.Drawing.Point(49,
129);
            this.label9.Name = "label9";
            this.label9.Size = new System.Drawing.Size(225,
18);
            this.label9.TabIndex = 40;
            this.label9.TextAlign = System.Drawing.
ContentAlignment.MiddleCenter;
            //
            // label10
            //
            this.label10.Font = new System.Drawing.
Font("Microsoft Sans Serif", 8.25F, System.Drawing.FontStyle.
Bold, System.Drawing.GraphicsUnit.Point, ((byte)(204)));
            this.label10.ForeColor = System.Drawing.Color.
LightCoral;
            this.label10.Location = new System.Drawing.
Point(278, 127);
            this.label10.Name = "label10";
            this.label10.Size = new System.Drawing.Size(14,
19);
            this.label10.TabIndex = 39;
            this.label10.Text = "x";
            this.label10.TextAlign = System.Drawing.
ContentAlignment.MiddleCenter;
            //
            // label11
            //
            this.label11.AutoSize = true;
            this.label11.ForeColor = System.Drawing.Color.
LightCoral;
```

```
            this.label11.Location = new System.Drawing.Point(5,
131);
            this.label11.Name = "label11";
            this.label11.Size = new System.Drawing.Size(44,
13);
            this.label11.TabIndex = 38;
            this.label11.Text = "AC N°3:";
            //
            // label6
            //
            this.label6.BorderStyle = System.Windows.Forms.
BorderStyle.FixedSingle;
            this.label6.Font = new System.Drawing.Font("Lucida
Console", 8.25F, System.Drawing.FontStyle.Regular, System.
Drawing.GraphicsUnit.Point, ((byte)(204)));
            this.label6.Location = new System.Drawing.Point(49,
94);
            this.label6.Name = "label6";
            this.label6.Size = new System.Drawing.Size(225,
18);
            this.label6.TabIndex = 37;
            this.label6.TextAlign = System.Drawing.
ContentAlignment.MiddleCenter;
            //
            // label7
            //
            this.label7.Font = new System.Drawing.
Font("Microsoft Sans Serif", 8.25F, System.Drawing.FontStyle.
Bold, System.Drawing.GraphicsUnit.Point, ((byte)(204)));
            this.label7.ForeColor = System.Drawing.Color.
OrangeRed;
            this.label7.Location = new System.Drawing.
Point(278, 92);
            this.label7.Name = "label7";
            this.label7.Size = new System.Drawing.Size(14, 19);
            this.label7.TabIndex = 36;
            this.label7.Text = "x";
            this.label7.TextAlign = System.Drawing.
ContentAlignment.MiddleCenter;
            //
            // label8
            //
            this.label8.AutoSize = true;
            this.label8.ForeColor = System.Drawing.Color.
OrangeRed;
            this.label8.Location = new System.Drawing.Point(5,
96);
            this.label8.Name = "label8";
            this.label8.Size = new System.Drawing.Size(44, 13);
```

```
            this.label8.TabIndex = 35;
            this.label8.Text = "AC N°2:";
            //
            // label5
            //
            this.label5.BorderStyle = System.Windows.Forms.
BorderStyle.FixedSingle;
            this.label5.Font = new System.Drawing.Font("Lucida
Console", 8.25F, System.Drawing.FontStyle.Regular, System.
Drawing.GraphicsUnit.Point, ((byte)(204)));
            this.label5.ForeColor = System.Drawing.
SystemColors.ControlText;
            this.label5.Location = new System.Drawing.Point(49,
59);
            this.label5.Name = "label5";
            this.label5.Size = new System.Drawing.Size(225,
18);
            this.label5.TabIndex = 34;
            this.label5.TextAlign = System.Drawing.
ContentAlignment.MiddleCenter;
            //
            // label4
            //
            this.label4.Font = new System.Drawing.
Font("Microsoft Sans Serif", 8.25F, System.Drawing.FontStyle.
Bold, System.Drawing.GraphicsUnit.Point, ((byte)(204)));
            this.label4.ForeColor = System.Drawing.Color.Red;
            this.label4.Location = new System.Drawing.
Point(278, 57);
            this.label4.Name = "label4";
            this.label4.Size = new System.Drawing.Size(14, 19);
            this.label4.TabIndex = 33;
            this.label4.Text = "x";
            this.label4.TextAlign = System.Drawing.
ContentAlignment.MiddleCenter;
            //
            // label3
            //
            this.label3.AutoSize = true;
            this.label3.ForeColor = System.Drawing.Color.Red;
            this.label3.Location = new System.Drawing.Point(5,
61);
            this.label3.Name = "label3";
            this.label3.Size = new System.Drawing.Size(44, 13);
            this.label3.TabIndex = 32;
            this.label3.Text = "AK N°1:";
            //
            // pictureBox2
            //
```

```
                this.pictureBox2.BackColor = System.Drawing.Color.
White;
                this.pictureBox2.BorderStyle = System.Windows.
Forms.BorderStyle.FixedSingle;
                this.pictureBox2.Location = new System.Drawing.
Point(17, 211);
                this.pictureBox2.Name = "pictureBox2";
                this.pictureBox2.Size = new System.Drawing.
Size(258, 258);
                this.pictureBox2.TabIndex = 31;
                this.pictureBox2.TabStop = false;
                //
                // label23
                //
                this.label23.AutoSize = true;
                this.label23.Location = new System.Drawing.
Point(239, 196);
                this.label23.Name = "label23";
                this.label23.Size = new System.Drawing.Size(13,
13);
                this.label23.TabIndex = 54;
                this.label23.Text = "0";
                //
                // label22
                //
                this.label22.AutoSize = true;
                this.label22.Location = new System.Drawing.
Point(200, 196);
                this.label22.Name = "label22";
                this.label22.Size = new System.Drawing.Size(33,
13);
                this.label22.TabIndex = 53;
                this.label22.Text = "byte:";
                //
                // label20
                //
                this.label20.AutoSize = true;
                this.label20.Location = new System.Drawing.
Point(148, 196);
                this.label20.Name = "label20";
                this.label20.Size = new System.Drawing.Size(13,
13);
                this.label20.TabIndex = 52;
                this.label20.Text = "0";
                //
                // label18
                //
                this.label18.AutoSize = true;
```

```
            this.label18.Location = new System.Drawing.
Point(46, 196);
            this.label18.Name = "label18";
            this.label18.Size = new System.Drawing.Size(96,
13);
            this.label18.TabIndex = 51;
            this.label18.Text = "Total iterations:";
            //
            // timer1
            //
            this.timer1.Tick += new System.EventHandler(this.
timer1_Tick);
            //
            // menuStrip1
            //
            this.menuStrip1.Items.AddRange(new System.Windows.
Forms.ToolStripItem[] {
            this.fileToolStripMenuItem,
            this.referenceToolStripMenuItem});
            this.menuStrip1.Location = new System.Drawing.
Point(0, 0);
            this.menuStrip1.Name = "menuStrip1";
            this.menuStrip1.Size = new System.Drawing.Size(624,
24);
            this.menuStrip1.TabIndex = 32;
            this.menuStrip1.Text = "menuStrip1";
            //
            // fileToolStripMenuItem
            //
            this.fileToolStripMenuItem.DropDownItems.
AddRange(new System.Windows.Forms.ToolStripItem[] {
            this.
saveConditionCAinExternalFileToolStripMenuItem,
            this.
loadConditionCAFromExternalFileToolStripMenuItem,
            this.toolStripSeparator1,
            this.aboutThePprogramToolStripMenuItem});
            this.fileToolStripMenuItem.Name =
"fileToolStripMenuItem";
            this.fileToolStripMenuItem.Size = new System.
Drawing.Size(48, 20);
            this.fileToolStripMenuItem.Text = "File";
            //
            // saveConditionCAinExternalFileToolStripMenuItem
            //
            this.
saveConditionCAinExternalFileToolStripMenuItem.Name =
"saveConditionCAinExternalFileToolStripMenuItem";
```

```
            this.
saveConditionCAinExternalFileToolStripMenuItem.Size = new
System.Drawing.Size(317, 22);
            this.
saveConditionCAinExternalFileToolStripMenuItem.Text = "Save CA
state to external file";
            this.
saveConditionCAinExternalFileToolStripMenuItem.
Click += new System.EventHandler(this.
saveConditionCAinExternalFileToolStripMenuItem_Click);
            //
            // loadConditionCAFromExternalFileToolStripMenuItem
            //
            this.
loadConditionCAFromExternalFileToolStripMenuItem.Name =
"loadConditionCAFromExternalFileToolStripMenuItem";
            this.
loadConditionCAFromExternalFileToolStripMenuItem.Size = new
System.Drawing.Size(317, 22);
            this.
loadConditionCAFromExternalFileToolStripMenuItem.Text = "load
Condition CA From External File";
            this.
loadConditionCAFromExternalFileToolStripMenuItem.
Click += new System.EventHandler(this.
loadConditionCAFromExternalFileToolStripMenuItem_Click);
            //
            // toolStripSeparator1
            //
            this.toolStripSeparator1.Name =
"toolStripSeparator1";
            this.toolStripSeparator1.Size = new System.Drawing.
Size(314, 6);
            //
            // ExitfromProgramToolStripMenuItem
            //
            this.exitFromProgramToolStripMenuItem.Name =
"exitFromProgramToolStripMenuItem";
            this.exitFromProgramToolStripMenuItem.Size = new
System.Drawing.Size(317, 22);
            this.exitFromProgramToolStripMenuItem.Text = "exit
From Program";
            this.exitFromProgramToolStripMenuItem.Click += new
System.EventHandler(this.exitFromProgramToolStripMenuItem_
Click);
            //
            // referenceToolStripMenuItem
            //
```

```
            this.referenceToolStripMenuItem.DropDownItems.
AddRange(new System.Windows.Forms.ToolStripItem[] {
            this.aboutTheProgramToolStripMenuItem});
            this.referenceToolStripMenuItem.Name =
"referenceToolStripMenuItem";
            this.referenceToolStripMenuItem.Size = new System.
Drawing.Size(65, 20);
            this.referenceToolStripMenuItem.Text = "Reference";
            //
            // aboutTheProgramToolStripMenuItem
            //
            this.aboutTheProgramToolStripMenuItem.Name =
"aboutTheProgramToolStripMenuItem";
            this.aboutTheProgramToolStripMenuItem.Size = new
System.Drawing.Size(158, 22);
            this.aboutTheProgramToolStripMenuItem.Text = "About
The Program...";
            this.aboutTheProgramToolStripMenuItem.Click += new
System.EventHandler(this.aboutTheProgramToolStripMenuItem_
Click);
            //
            // openFileDialog1
            //
            this.openFileDialog1.FileName = "openFileDialog1";
            //
            // button3
            //
            this.button3.Location = new System.Drawing.
Point(484, 1);
            this.button3.Name = "button3";
            this.button3.Size = new System.Drawing.Size(140,
23);
            this.button3.TabIndex = 33;
            this.button3.Text = "Hide CA field";
            this.button3.UseVisualStyleBackColor = true;
            this.button3.Click += new System.EventHandler(this.
button3_Click);
            //
            // tabPage8
            //
            this.tabPage8.Controls.Add(this.pictureBox6);
            this.tabPage8.Controls.Add(this.pictureBox7);
            this.tabPage8.Controls.Add(this.label41);
            this.tabPage8.Controls.Add(this.label42);
            this.tabPage8.Controls.Add(this.label43);
            this.tabPage8.Controls.Add(this.label44);
            this.tabPage8.Controls.Add(this.label45);
            this.tabPage8.Controls.Add(this.label46);
            this.tabPage8.Controls.Add(this.label47);
```

```
            this.tabPage8.Location = new System.Drawing.
Point(4, 22);
            this.tabPage8.Name = "tabPage8";
            this.tabPage8.Padding = new System.Windows.Forms.
Padding(3);
            this.tabPage8.Size = new System.Drawing.Size(319,
580);
            this.tabPage8.TabIndex = 0;
            this.tabPage8.Text = "AK N°1";
            this.tabPage8.UseVisualStyleBackColor = true;
            //
```

// pictureBox6

```
            //
            this.pictureBox6.BackColor = System.Drawing.Color.
White;
            this.pictureBox6.BorderStyle = System.Windows.
Forms.BorderStyle.FixedSingle;
            this.pictureBox6.Location = new System.Drawing.
Point(17, 211);
            this.pictureBox6.Name = "pictureBox6";
            this.pictureBox6.Size = new System.Drawing.
Size(258, 258);
            this.pictureBox6.TabIndex = 59;
            this.pictureBox6.TabStop = false;
            //
            // pictureBox7
            //
            this.pictureBox7.BackColor = System.Drawing.Color.
White;
            this.pictureBox7.BorderStyle = System.Windows.
Forms.BorderStyle.FixedSingle;
            this.pictureBox7.Location = new System.Drawing.
Point(16, 24);
            this.pictureBox7.Name = "pictureBox7";
            this.pictureBox7.Size = new System.Drawing.
Size(258, 140);
            this.pictureBox7.SizeMode = System.Windows.Forms.
PictureBoxSizeMode.StretchImage;
            this.pictureBox7.TabIndex = 64;
            this.pictureBox7.TabStop = false;
            //
            // label41
            //
            this.label41.AutoSize = true;
            this.label41.ForeColor = System.Drawing.
SystemColors.ControlText;
            this.label41.Location = new System.Drawing.
Point(14, 470);
            this.label41.Name = "label41";
```

```
            this.label41.Size = new System.Drawing.Size(34,
13);
            this.label41.TabIndex = 63;
            this.label41.Text = "FF,00";
            //
            // label42
            //
            this.label42.AutoSize = true;
            this.label42.ForeColor = System.Drawing.
SystemColors.ControlText;
            this.label42.Location = new System.Drawing.
Point(14, 8);
            this.label42.Name = "label42";
            this.label42.Size = new System.Drawing.Size(23,
13);
            this.label42.TabIndex = 67;
            this.label42.Text = "pcs";
            //
            // label43
            //
            this.label43.AutoSize = true;
            this.label43.ForeColor = System.Drawing.
SystemColors.ControlText;
            this.label43.Location = new System.Drawing.
Point(249, 470);
            this.label43.Name = "label43";
            this.label43.Size = new System.Drawing.Size(34,
13);
            this.label43.TabIndex = 60;
            this.label43.Text = "FF,FF";
            //
            // label44
            //
            this.label44.AutoSize = true;
            this.label44.ForeColor = System.Drawing.
SystemColors.ControlText;
            this.label44.Location = new System.Drawing.
Point(13, 167);
            this.label44.Name = "label44";
            this.label44.Size = new System.Drawing.Size(13,
13);
            this.label44.TabIndex = 65;
            this.label44.Text = "0";
            //
            // label45
            //
            this.label45.AutoSize = true;
            this.label45.ForeColor = System.Drawing.
SystemColors.ControlText;
```

```
            this.label45.Location = new System.Drawing.
Point(249, 195);
            this.label45.Name = "label45";
            this.label45.Size = new System.Drawing.Size(34,
13);
            this.label45.TabIndex = 62;
            this.label45.Text = "00,FF";
            //
            // label46
            //
            this.label46.AutoSize = true;
            this.label46.ForeColor = System.Drawing.
SystemColors.ControlText;
            this.label46.Location = new System.Drawing.
Point(236, 167);
            this.label46.Name = "label46";
            this.label46.Size = new System.Drawing.Size(38,
13);
            this.label46.TabIndex = 66;
            this.label46.Text = " bytes";
                                   //
            // label47
            //
            this.label47.AutoSize = true;
            this.label47.ForeColor = System.Drawing.
SystemColors.ControlText;
            this.label47.Location = new System.Drawing.
Point(14, 195);
            this.label47.Name = "label47";
            this.label47.Size = new System.Drawing.Size(34,
13);
            this.label47.TabIndex = 61;
            this.label47.Text = "00,00";
            //
            // tabPage9
            //
            this.tabPage9.Controls.Add(this.pictureBox8);
            this.tabPage9.Controls.Add(this.pictureBox9);
            this.tabPage9.Controls.Add(this.label48);
            this.tabPage9.Controls.Add(this.label49);
            this.tabPage9.Controls.Add(this.label50);
            this.tabPage9.Controls.Add(this.label51);
            this.tabPage9.Controls.Add(this.label52);
            this.tabPage9.Controls.Add(this.label53);
            this.tabPage9.Controls.Add(this.label54);
            this.tabPage9.Location = new System.Drawing.
Point(4, 22);
            this.tabPage9.Name = "tabPage9";
```

```
            this.tabPage9.Padding = new System.Windows.Forms.
Padding(3);
            this.tabPage9.Size = new System.Drawing.Size(319,
580);
            this.tabPage9.TabIndex = 1;
            this.tabPage9.Text = "AK N°2";
            this.tabPage9.UseVisualStyleBackColor = true;
            //
            // pictureBox8
            //
            this.pictureBox8.BackColor = System.Drawing.Color.
White;
            this.pictureBox8.BorderStyle = System.Windows.
Forms.BorderStyle.FixedSingle;
            this.pictureBox8.Location = new System.Drawing.
Point(17, 211);
            this.pictureBox8.Name = "pictureBox8";
            this.pictureBox8.Size = new System.Drawing.
Size(258, 258);
            this.pictureBox8.TabIndex = 68;
            this.pictureBox8.TabStop = false;
            //
            // pictureBox9
            //
            this.pictureBox9.BackColor = System.Drawing.Color.
White;
            this.pictureBox9.BorderStyle = System.Windows.
Forms.BorderStyle.FixedSingle;
            this.pictureBox9.Location = new System.Drawing.
Point(16, 24);
            this.pictureBox9.Name = "pictureBox9";
            this.pictureBox9.Size = new System.Drawing.
Size(258, 140);
            this.pictureBox9.SizeMode = System.Windows.Forms.
PictureBoxSizeMode.StretchImage;
            this.pictureBox9.TabIndex = 73;
            this.pictureBox9.TabStop = false;
            //
            // label48
            //
            this.label48.AutoSize = true;
            this.label48.ForeColor = System.Drawing.
SystemColors.ControlText;
            this.label48.Location = new System.Drawing.
Point(14, 470);
            this.label48.Name = "label48";
            this.label48.Size = new System.Drawing.Size(34,
13);
            this.label48.TabIndex = 72;
```

```
            this.label48.Text = "FF,00";
            //
            // label49
            //
            this.label49.AutoSize = true;
            this.label49.ForeColor = System.Drawing.
SystemColors.ControlText;
            this.label49.Location = new System.Drawing.
Point(14, 8);
            this.label49.Name = "label49";
            this.label49.Size = new System.Drawing.Size(23,
13);
            this.label49.TabIndex = 76;
            this.label49.Text = "pcs";
            //
            // label50
                                          //
            this.label50.AutoSize = true;
            this.label50.ForeColor = System.Drawing.
SystemColors.ControlText;
            this.label50.Location = new System.Drawing.
Point(249, 470);
            this.label50.Name = "label50";
            this.label50.Size = new System.Drawing.Size(34,
13);
            this.label50.TabIndex = 69;
            this.label50.Text = "FF,FF";
            //
            // label51
            //
            this.label51.AutoSize = true;
            this.label51.ForeColor = System.Drawing.
SystemColors.ControlText;
            this.label51.Location = new System.Drawing.
Point(13, 167);
            this.label51.Name = "label51";
            this.label51.Size = new System.Drawing.Size(13,
13);
            this.label51.TabIndex = 74;
            this.label51.Text = "0";
            //
            // label52
            //
            this.label52.AutoSize = true;
            this.label52.ForeColor = System.Drawing.
SystemColors.ControlText;
            this.label52.Location = new System.Drawing.
Point(249, 195);
            this.label52.Name = "label52";
```

```
            this.label52.Size = new System.Drawing.Size(34,
13);
            this.label52.TabIndex = 71;
            this.label52.Text = "00,FF";
            //
            // label53
            //
            this.label53.AutoSize = true;
            this.label53.ForeColor = System.Drawing.
SystemColors.ControlText;
            this.label53.Location = new System.Drawing.
Point(236, 167);
            this.label53.Name = "label53";
            this.label53.Size = new System.Drawing.Size(38,
13);
            this.label53.TabIndex = 75;
            this.label53.Text = "bytes";
            //
            // label54
            //
            this.label54.AutoSize = true;
            this.label54.ForeColor = System.Drawing.
SystemColors.ControlText;
            this.label54.Location = new System.Drawing.
Point(14, 195);
            this.label54.Name = "label54";
            this.label54.Size = new System.Drawing.Size(34,
13);
            this.label54.TabIndex = 70;
            this.label54.Text = "00,00";
            //
            // tabPage2
            //
            this.tabPage2.BackColor = System.Drawing.
SystemColors.Control;
            this.tabPage2.Controls.Add(this.tabControl2);
            this.tabPage2.Location = new System.Drawing.
Point(4, 22);
            this.tabPage2.Name = "tabPage2";
            this.tabPage2.Padding = new System.Windows.Forms.
Padding(3);
            this.tabPage2.Size = new System.Drawing.Size(317,
532);
            this.tabPage2.TabIndex = 1;
            this.tabPage2.Text = "Graphical representation of
the results";
            //
            // tabControl2
            //
```

```
            this.tabControl2.Controls.Add(this.tabPage11);
            this.tabControl2.Controls.Add(this.tabPage6);
            this.tabControl2.Controls.Add(this.tabPage7);
            this.tabControl2.Controls.Add(this.tabPage10);
            this.tabControl2.Location = new System.Drawing.
Point(3, 6);
            this.tabControl2.Name = "tabControl2";
            this.tabControl2.SelectedIndex = 0;
            this.tabControl2.Size = new System.Drawing.
Size(311, 521);
            this.tabControl2.TabIndex = 0;
            //
            // tabPage11
            //
            this.tabPage11.BackColor = System.Drawing.
SystemColors.Control;
            this.tabPage11.Controls.Add(this.pictureBox2);
            this.tabPage11.Controls.Add(this.pictureBox1);
            this.tabPage11.Controls.Add(this.label62);
            this.tabPage11.Controls.Add(this.label63);
            this.tabPage11.Controls.Add(this.label64);
            this.tabPage11.Controls.Add(this.label65);
            this.tabPage11.Controls.Add(this.label66);
            this.tabPage11.Controls.Add(this.label67);
            this.tabPage11.Controls.Add(this.label68);
            this.tabPage11.Location = new System.Drawing.
Point(4, 22);
            this.tabPage11.Name = "tabPage11";
            this.tabPage11.Padding = new System.Windows.Forms.
Padding(3);
            this.tabPage11.Size = new System.Drawing.Size(303,
495);
            this.tabPage11.TabIndex = 3;
            this.tabPage11.Text = "XOR";
            //
            // pictureBox1
            //
            this.pictureBox1.BackColor = System.Drawing.Color.
White;
            this.pictureBox1.BorderStyle = System.Windows.
Forms.BorderStyle.FixedSingle;
            this.pictureBox1.Location = new System.Drawing.
Point(16, 24);
            this.pictureBox1.Name = "pictureBox1";
            this.pictureBox1.Size = new System.Drawing.
Size(258, 142);
            this.pictureBox1.SizeMode = System.Windows.Forms.
PictureBoxSizeMode.StretchImage;
            this.pictureBox1.TabIndex = 73;
```

```
            this.pictureBox1.TabStop = false;
            //
            // label62
            //
            this.label62.AutoSize = true;
            this.label62.ForeColor = System.Drawing.
SystemColors.ControlText;
            this.label62.Location = new System.Drawing.
Point(249, 195);
            this.label62.Name = "label62";
            this.label62.Size = new System.Drawing.Size(34,
13);
            this.label62.TabIndex = 72;
            this.label62.Text = "FF,00";
            //
            // label63
            //
            this.label63.AutoSize = true;
            this.label63.ForeColor = System.Drawing.
SystemColors.ControlText;
            this.label63.Location = new System.Drawing.
Point(14, 8);
            this.label63.Name = "label63";
            this.label63.Size = new System.Drawing.Size(23,
13);
            this.label63.TabIndex = 76;
            this.label63.Text = "pcs";
            //
            // label64
            //
            this.label64.AutoSize = true;
            this.label64.ForeColor = System.Drawing.
SystemColors.ControlText;
            this.label64.Location = new System.Drawing.
Point(249, 470);
            this.label64.Name = "label64";
            this.label64.Size = new System.Drawing.Size(34,
13);
            this.label64.TabIndex = 69;
            this.label64.Text = "FF,FF";
            //
            // label65
            //
            this.label65.AutoSize = true;
            this.label65.ForeColor = System.Drawing.
SystemColors.ControlText;
            this.label65.Location = new System.Drawing.
Point(13, 167);
            this.label65.Name = "label65";
```

```
            this.label65.Size = new System.Drawing.Size(13,
13);
            this.label65.TabIndex = 74;
            this.label65.Text = "0";
            //
            // label66
            //
            this.label66.AutoSize = true;
            this.label66.ForeColor = System.Drawing.
SystemColors.ControlText;
            this.label66.Location = new System.Drawing.
Point(14, 470);
            this.label66.Name = "label66";
            this.label66.Size = new System.Drawing.Size(34,
13);
            this.label66.TabIndex = 71;
            this.label66.Text = "00,FF";
            //
            // label67
            //
            this.label67.AutoSize = true;
            this.label67.ForeColor = System.Drawing.
SystemColors.ControlText;
            this.label67.Location = new System.Drawing.
Point(236, 167);
            this.label67.Name = "label67";
            this.label67.Size = new System.Drawing.Size(38,
13);
            this.label67.TabIndex = 75;
            this.label67.Text = "bytes";
            //
            // label68
            //
            this.label68.AutoSize = true;
            this.label68.ForeColor = System.Drawing.
SystemColors.ControlText;
            this.label68.Location = new System.Drawing.
Point(14, 195);
            this.label68.Name = "label68";
            this.label68.Size = new System.Drawing.Size(34,
13);
            this.label68.TabIndex = 70;
            this.label68.Text = "00,00";
            //
            // tabPage6
            //
            this.tabPage6.BackColor = System.Drawing.
SystemColors.Control;
            this.tabPage6.Controls.Add(this.pictureBoxAll);
```

```
            this.tabPage6.Controls.Add(this.pictureBoxA12);
            this.tabPage6.Controls.Add(this.label30);
            this.tabPage6.Controls.Add(this.label27);
            this.tabPage6.Controls.Add(this.label31);
            this.tabPage6.Controls.Add(this.label29);
            this.tabPage6.Controls.Add(this.label32);
            this.tabPage6.Controls.Add(this.label28);
            this.tabPage6.Controls.Add(this.label33);
            this.tabPage6.ForeColor = System.Drawing.
SystemColors.ControlText;
            this.tabPage6.Location = new System.Drawing.
Point(4, 22);
            this.tabPage6.Name = "tabPage6";
            this.tabPage6.Padding = new System.Windows.Forms.
Padding(3);
            this.tabPage6.Size = new System.Drawing.Size(303,
495);
            this.tabPage6.TabIndex = 0;
            this.tabPage6.Text = "AC N°1";
            //
            // pictureBoxA11
            //
            this.pictureBoxA11.BackColor = System.Drawing.
Color.White;
            this.pictureBoxA11.BorderStyle = System.Windows.
Forms.BorderStyle.FixedSingle;
            this.pictureBoxA11.Location = new System.Drawing.
Point(16, 24);
            this.pictureBoxA11.Name = "pictureBoxA11";
            this.pictureBoxA11.Size = new System.Drawing.
Size(258, 142);
            this.pictureBoxA11.SizeMode = System.Windows.Forms.
PictureBoxSizeMode.StretchImage;
            this.pictureBoxA11.TabIndex = 74;
            this.pictureBoxA11.TabStop = false;
            //
            // pictureBoxA12
            //
            this.pictureBoxA12.BackColor = System.Drawing.
Color.White;
            this.pictureBoxA12.BorderStyle = System.Windows.
Forms.BorderStyle.FixedSingle;
            this.pictureBoxA12.Location = new System.Drawing.
Point(17, 211);
            this.pictureBoxA12.Name = "pictureBoxA12";
            this.pictureBoxA12.Size = new System.Drawing.
Size(258, 258);
            this.pictureBoxA12.TabIndex = 59;
            this.pictureBoxA12.TabStop = false;
```

```
            //
            // label30
            //
            this.label30.AutoSize = true;
            this.label30.ForeColor = System.Drawing.
SystemColors.ControlText;
            this.label30.Location = new System.Drawing.
Point(249, 195);
            this.label30.Name = "label30";
            this.label30.Size = new System.Drawing.Size(34,
13);
            this.label30.TabIndex = 63;
            this.label30.Text = "FF,00";
            //
            // label27
            //
            this.label27.AutoSize = true;
            this.label27.ForeColor = System.Drawing.
SystemColors.ControlText;
            this.label27.Location = new System.Drawing.
Point(14, 8);
            this.label27.Name = "label27";
            this.label27.Size = new System.Drawing.Size(23,
13);
            this.label27.TabIndex = 67;
            this.label27.Text = "pcs";
            //
            // label31
            //
            this.label31.AutoSize = true;
            this.label31.ForeColor = System.Drawing.
SystemColors.ControlText;
            this.label31.Location = new System.Drawing.
Point(249, 470);
            this.label31.Name = "label31";
            this.label31.Size = new System.Drawing.Size(34,
13);
            this.label31.TabIndex = 60;
            this.label31.Text = "FF,FF";
            //
            // label29
            //
            this.label29.AutoSize = true;
            this.label29.ForeColor = System.Drawing.
SystemColors.ControlText;
            this.label29.Location = new System.Drawing.
Point(13, 167);
            this.label29.Name = "label29";
```

```
                this.label29.Size = new System.Drawing.Size(13,
13);
                this.label29.TabIndex = 65;
                this.label29.Text = "0";
                //
                // label32
                //
                this.label32.AutoSize = true;
                this.label32.ForeColor = System.Drawing.
SystemColors.ControlText;
                this.label32.Location = new System.Drawing.
Point(14, 470);
                this.label32.Name = "label32";
                this.label32.Size = new System.Drawing.Size(34,
13);
                this.label32.TabIndex = 62;
                this.label32.Text = "00,FF";
                //
                // label28
                //
                this.label28.AutoSize = true;
                this.label28.ForeColor = System.Drawing.
SystemColors.ControlText;
                this.label28.Location = new System.Drawing.
Point(236, 167);
                this.label28.Name = "label28";
                this.label28.Size = new System.Drawing.Size(38,
13);
                this.label28.TabIndex = 66;
                this.label28.Text = "bytes";
                //
                // label33
                //
                this.label33.AutoSize = true;
                this.label33.ForeColor = System.Drawing.
SystemColors.ControlText;
                this.label33.Location = new System.Drawing.
Point(14, 195);
                this.label33.Name = "label33";
                this.label33.Size = new System.Drawing.Size(34,
13);
                this.label33.TabIndex = 61;
                this.label33.Text = "00,00";
                //
                // tabPage7
                //
                this.tabPage7.BackColor = System.Drawing.
SystemColors.Control;
                this.tabPage7.Controls.Add(this.pictureBoxA22);
```

```
            this.tabPage7.Controls.Add(this.pictureBoxA21);
            this.tabPage7.Controls.Add(this.label34);
            this.tabPage7.Controls.Add(this.label35);
            this.tabPage7.Controls.Add(this.label36);
            this.tabPage7.Controls.Add(this.label37);
            this.tabPage7.Controls.Add(this.label38);
            this.tabPage7.Controls.Add(this.label39);
            this.tabPage7.Controls.Add(this.label40);
            this.tabPage7.Location = new System.Drawing.
Point(4, 22);
            this.tabPage7.Name = "tabPage7";
            this.tabPage7.Padding = new System.Windows.Forms.
Padding(3);
            this.tabPage7.Size = new System.Drawing.Size(303,
495);
            this.tabPage7.TabIndex = 1;
            this.tabPage7.Text = "AC N°2";
            //
            // pictureBoxA22
            //
            this.pictureBoxA22.BackColor = System.Drawing.
Color.White;
            this.pictureBoxA22.BorderStyle = System.Windows.
Forms.BorderStyle.FixedSingle;
            this.pictureBoxA22.Location = new System.Drawing.
Point(17, 211);
            this.pictureBoxA22.Name = "pictureBoxA22";
            this.pictureBoxA22.Size = new System.Drawing.
Size(258, 258);
            this.pictureBoxA22.TabIndex = 68;
            this.pictureBoxA22.TabStop = false;
            //
            // pictureBoxA21
            //
            this.pictureBoxA21.BackColor = System.Drawing.
Color.White;
            this.pictureBoxA21.BorderStyle = System.Windows.
Forms.BorderStyle.FixedSingle;
            this.pictureBoxA21.Location = new System.Drawing.
Point(16, 24);
            this.pictureBoxA21.Name = "pictureBoxA21";
            this.pictureBoxA21.Size = new System.Drawing.
Size(258, 142);
            this.pictureBoxA21.SizeMode = System.Windows.Forms.
PictureBoxSizeMode.StretchImage;
            this.pictureBoxA21.TabIndex = 73;
            this.pictureBoxA21.TabStop = false;
            //
            // label34
```

```
            //
            this.label34.AutoSize = true;
            this.label34.ForeColor = System.Drawing.
SystemColors.ControlText;
            this.label34.Location = new System.Drawing.
Point(249, 195);
            this.label34.Name = "label34";
            this.label34.Size = new System.Drawing.Size(34,
13);
            this.label34.TabIndex = 72;
            this.label34.Text = "FF,00";
            //
            // label35
            //
            this.label35.AutoSize = true;
            this.label35.ForeColor = System.Drawing.
SystemColors.ControlText;
            this.label35.Location = new System.Drawing.
Point(14, 8);
            this.label35.Name = "label35";
            this.label35.Size = new System.Drawing.Size(23,
13);
            this.label35.TabIndex = 76;
            this.label35.Text = "pcs";
            //
            // label36
            //
            this.label36.AutoSize = true;
            this.label36.ForeColor = System.Drawing.
SystemColors.ControlText;
            this.label36.Location = new System.Drawing.
Point(249, 470);
            this.label36.Name = "label36";
            this.label36.Size = new System.Drawing.Size(34,
13);
            this.label36.TabIndex = 69;
            this.label36.Text = "FF,FF";
            //
            // label37
            //
            this.label37.AutoSize = true;
            this.label37.ForeColor = System.Drawing.
SystemColors.ControlText;
            this.label37.Location = new System.Drawing.
Point(13, 167);
            this.label37.Name = "label37";
            this.label37.Size = new System.Drawing.Size(13,
13);
            this.label37.TabIndex = 74;
```

```
            this.label37.Text = "0";
            //
            // label38
            //
            this.label38.AutoSize = true;
            this.label38.ForeColor = System.Drawing.
SystemColors.ControlText;
            this.label38.Location = new System.Drawing.
Point(14, 470);
            this.label38.Name = "label38";
            this.label38.Size = new System.Drawing.Size(34,
13);
            this.label38.TabIndex = 71;
            this.label38.Text = "00,FF";
            //
            // label39
            //
            this.label39.AutoSize = true;
            this.label39.ForeColor = System.Drawing.
SystemColors.ControlText;
            this.label39.Location = new System.Drawing.
Point(236, 167);
            this.label39.Name = "label39";
            this.label39.Size = new System.Drawing.Size(38,
13);
            this.label39.TabIndex = 75;
            this.label39.Text = "bytes";
            //
            // label40
            //
            this.label40.AutoSize = true;
            this.label40.ForeColor = System.Drawing.
SystemColors.ControlText;
            this.label40.Location = new System.Drawing.
Point(14, 195);
            this.label40.Name = "label40";
            this.label40.Size = new System.Drawing.Size(34,
13);
            this.label40.TabIndex = 70;
            this.label40.Text = "00,00";
            //
            // tabPage10
            //
            this.tabPage10.BackColor = System.Drawing.
SystemColors.Control;
            this.tabPage10.Controls.Add(this.pictureBoxA32);
            this.tabPage10.Controls.Add(this.pictureBoxA31);
            this.tabPage10.Controls.Add(this.label55);
            this.tabPage10.Controls.Add(this.label56);
```

```
        this.tabPage10.Controls.Add(this.label57);
        this.tabPage10.Controls.Add(this.label58);
        this.tabPage10.Controls.Add(this.label59);
        this.tabPage10.Controls.Add(this.label60);
        this.tabPage10.Controls.Add(this.label61);
        this.tabPage10.Location = new System.Drawing.
Point(4, 22);
        this.tabPage10.Name = "tabPage10";
        this.tabPage10.Padding = new System.Windows.Forms.
Padding(3);
        this.tabPage10.Size = new System.Drawing.Size(303,
495);
        this.tabPage10.TabIndex = 2;
        this.tabPage10.Text = "AC N°3";
        //
        // pictureBoxA32
        //
        this.pictureBoxA32.BackColor = System.Drawing.
Color.White;
        this.pictureBoxA32.BorderStyle = System.Windows.
Forms.BorderStyle.FixedSingle;
        this.pictureBoxA32.Location = new System.Drawing.
Point(17, 211);
        this.pictureBoxA32.Name = "pictureBoxA32";
        this.pictureBoxA32.Size = new System.Drawing.
Size(258, 258);
        this.pictureBoxA32.TabIndex = 68;
        this.pictureBoxA32.TabStop = false;
        //
        // pictureBoxA31
        //
        this.pictureBoxA31.BackColor = System.Drawing.
Color.White;
        this.pictureBoxA31.BorderStyle = System.Windows.
Forms.BorderStyle.FixedSingle;
        this.pictureBoxA31.Location = new System.Drawing.
Point(16, 24);
        this.pictureBoxA31.Name = "pictureBoxA31";
        this.pictureBoxA31.Size = new System.Drawing.
Size(258, 142);
        this.pictureBoxA31.SizeMode = System.Windows.Forms.
PictureBoxSizeMode.StretchImage;
        this.pictureBoxA31.TabIndex = 73;
        this.pictureBoxA31.TabStop = false;
        //
        // label55
        //
        this.label55.AutoSize = true;
```

```
            this.label55.ForeColor = System.Drawing.
SystemColors.ControlText;
            this.label55.Location = new System.Drawing.
Point(249, 195);
            this.label55.Name = "label55";
            this.label55.Size = new System.Drawing.Size(34,
13);
            this.label55.TabIndex = 72;
            this.label55.Text = "FF,00";
            //
            // label56
            //
            this.label56.AutoSize = true;
            this.label56.ForeColor = System.Drawing.
SystemColors.ControlText;
            this.label56.Location = new System.Drawing.
Point(14, 8);
            this.label56.Name = "label56";
            this.label56.Size = new System.Drawing.Size(23,
13);
            this.label56.TabIndex = 76;
            this.label56.Text = "pcs";
            //
            // label57
            //
            this.label57.AutoSize = true;
            this.label57.ForeColor = System.Drawing.
SystemColors.ControlText;
            this.label57.Location = new System.Drawing.
Point(249, 470);
            this.label57.Name = "label57";
            this.label57.Size = new System.Drawing.Size(34,
13);
            this.label57.TabIndex = 69;
            this.label57.Text = "FF,FF";
            //
            // label58
            //
            this.label58.AutoSize = true;
            this.label58.ForeColor = System.Drawing.
SystemColors.ControlText;
            this.label58.Location = new System.Drawing.
Point(13, 167);
            this.label58.Name = "label58";
            this.label58.Size = new System.Drawing.Size(13,
13);
            this.label58.TabIndex = 74;
            this.label58.Text = "0";
            //
```

```
// label59
//
this.label59.AutoSize = true;
this.label59.ForeColor = System.Drawing.
SystemColors.ControlText;
this.label59.Location = new System.Drawing.
Point(14, 470);
this.label59.Name = "label59";
this.label59.Size = new System.Drawing.Size(34,
13);
this.label59.TabIndex = 71;
this.label59.Text = "00,FF";
//
// label60
//
this.label60.AutoSize = true;
this.label60.ForeColor = System.Drawing.
SystemColors.ControlText;
this.label60.Location = new System.Drawing.
Point(236, 167);
this.label60.Name = "label60";
this.label60.Size = new System.Drawing.Size(38,
13);
this.label60.TabIndex = 75;
this.label60.Text = "bytes";
//
// label61
//
this.label61.AutoSize = true;
this.label61.ForeColor = System.Drawing.
SystemColors.ControlText;
this.label61.Location = new System.Drawing.
Point(14, 195);
this.label61.Name = "label61";
this.label61.Size = new System.Drawing.Size(34,
13);
this.label61.TabIndex = 70;
this.label61.Text = "00,00";
//
// tabPage1
//
this.tabPage1.BackColor = System.Drawing.
SystemColors.Control;
this.tabPage1.Controls.Add(this.groupBox6);
this.tabPage1.Controls.Add(this.groupBox4);
this.tabPage1.Controls.Add(this.groupBox2);
this.tabPage1.Location = new System.Drawing.
Point(4, 22);
this.tabPage1.Name = "tabPage1";
```

```
            this.tabPage1.Padding = new System.Windows.Forms.
Padding(3);
            this.tabPage1.Size = new System.Drawing.Size(317,
532);
            this.tabPage1.TabIndex = 0;
            this.tabPage1.Text = "CA control";
            //
            // groupBox6
            //
            this.groupBox6.Controls.Add(this.label23);
            this.groupBox6.Controls.Add(this.label3);
            this.groupBox6.Controls.Add(this.label22);
            this.groupBox6.Controls.Add(this.label16);
            this.groupBox6.Controls.Add(this.label9);
            this.groupBox6.Controls.Add(this.label20);
            this.groupBox6.Controls.Add(this.label14);
            this.groupBox6.Controls.Add(this.label18);
            this.groupBox6.Controls.Add(this.label10);
            this.groupBox6.Controls.Add(this.label13);
            this.groupBox6.Controls.Add(this.label11);
            this.groupBox6.Controls.Add(this.label6);
            this.groupBox6.Controls.Add(this.label4);
            this.groupBox6.Controls.Add(this.label17);
            this.groupBox6.Controls.Add(this.label5);
            this.groupBox6.Controls.Add(this.label7);
            this.groupBox6.Controls.Add(this.label15);
            this.groupBox6.Controls.Add(this.label8);
            this.groupBox6.Location = new System.Drawing.
Point(8, 296);
            this.groupBox6.Name = "groupBox6";
            this.groupBox6.Size = new System.Drawing.Size(300,
221);
            this.groupBox6.TabIndex = 30;
            this.groupBox6.TabStop = false;
            this.groupBox6.Text = "Generated Byte Sequences";
            //
            // tabControl1
            //
            this.tabControl1.Controls.Add(this.tabPage1);
            this.tabControl1.Controls.Add(this.tabPage2);
            this.tabControl1.Location = new System.Drawing.
Point(3, 30);
            this.tabControl1.Name = "tabControl1";
            this.tabControl1.SelectedIndex = 0;
            this.tabControl1.Size = new System.Drawing.
Size(325, 558);
            this.tabControl1.TabIndex = 34;
            //
            // progressBar1
```

```
            //
            this.progressBar1.Location = new System.Drawing.
Point(4, 38);
            this.progressBar1.Name = "progressBar1";
            this.progressBar1.Size = new System.Drawing.
Size(277, 38);
            this.progressBar1.Step = 1;
            this.progressBar1.Style = System.Windows.Forms.
ProgressBarStyle.Continuous;
            this.progressBar1.TabIndex = 35;
            //
            // timer2
            //
            this.timer2.Interval = 500;
            this.timer2.Tick += new System.EventHandler(this.
timer2_Tick);
            //
            // panel2
            //
            this.panel2.Controls.Add(this.label75);
            this.panel2.Controls.Add(this.label73);
            this.panel2.Controls.Add(this.label72);
            this.panel2.Controls.Add(this.label74);
            this.panel2.Controls.Add(this.label69);
            this.panel2.Controls.Add(this.label70);
            this.panel2.Controls.Add(this.label71);
            this.panel2.Controls.Add(this.label26);
            this.panel2.Controls.Add(this.label25);
            this.panel2.Controls.Add(this.label24);
            this.panel2.Controls.Add(this.label12);
            this.panel2.Controls.Add(this.progressBar1);
            this.panel2.Location = new System.Drawing.
Point(333, 216);
            this.panel2.Name = "panel2";
            this.panel2.Size = new System.Drawing.Size(284,
211);
            this.panel2.TabIndex = 36;
            this.panel2.Visible = false;
            //
            // label75
            //
            this.label75.AutoSize = true;
            this.label75.Location = new System.Drawing.
Point(253, 151);
            this.label75.Name = "label75";
            this.label75.Size = new System.Drawing.Size(13,
13);
            this.label75.TabIndex = 47;
            this.label75.Text = "0";
```

```
                //
                // label73
                //
                this.label73.AutoSize = true;
                this.label73.Location = new System.Drawing.
Point(151, 151);
                this.label73.Name = "label73";
                this.label73.Size = new System.Drawing.Size(13,
13);
                this.label73.TabIndex = 46;
                this.label73.Text = "0";
                //
                // label72
                //
                this.label72.AutoSize = true;
                this.label72.Location = new System.Drawing.
Point(180, 151);
                this.label72.Name = "label72";
                this.label72.Size = new System.Drawing.Size(57,
13);
                this.label72.TabIndex = 45;
                this.label72.Text = "left:";
                //
                // label74
                //
                this.label74.Location = new System.Drawing.
Point(15, 151);
                this.label74.Name = "label74";
                this.label74.Size = new System.Drawing.Size(158,
17);
                this.label74.TabIndex = 43;
                this.label74.Text = "time has passed (min.):";
                //
                // label69
                //
                this.label69.AutoSize = true;
                this.label69.Location = new System.Drawing.
Point(196, 127);
                this.label69.Name = "label69";
                this.label69.Size = new System.Drawing.Size(41,
13);
                this.label69.TabIndex = 42;
                this.label69.Text = "byte/s";
                //
                // label70
                //
                this.label70.Location = new System.Drawing.
Point(126, 127);
                this.label70.Name = "label70";
```

```
            this.label70.Size = new System.Drawing.Size(69,
13);
            this.label70.TabIndex = 41;
            this.label70.Text = "0";
            this.label70.TextAlign = System.Drawing.
ContentAlignment.MiddleRight;
            //
            // label71
            //
            this.label71.AutoSize = true;
            this.label71.Location = new System.Drawing.
Point(15, 127);
            this.label71.Name = "label71";
            this.label71.Size = new System.Drawing.Size(113,
13);
            this.label71.TabIndex = 40;
            this.label71.Text = "generation rate:";
            //
            // label26
            //
            this.label26.AutoSize = true;
            this.label26.Location = new System.Drawing.
Point(196, 104);
            this.label26.Name = "label26";
            this.label26.Size = new System.Drawing.Size(19,
13);
            this.label26.TabIndex = 39;
            this.label26.Text = "from";
            //
            // label25
            //
            this.label25.AutoSize = true;
            this.label25.Location = new System.Drawing.
Point(217, 104);
            this.label25.Name = "label25";
            this.label25.Size = new System.Drawing.Size(13,
13);
            this.label25.TabIndex = 38;
            this.label25.Text = "0";
            //
            // label24
            //
            this.label24.Location = new System.Drawing.
Point(126, 104);
            this.label24.Name = "label24";
            this.label24.Size = new System.Drawing.Size(69,
13);
            this.label24.TabIndex = 37;
            this.label24.Text = "0";
```

```
            this.label24.TextAlign = System.Drawing.
ContentAlignment.MiddleRight;
            //
            // label12
            //
            this.label12.AutoSize = true;
            this.label12.Location = new System.Drawing.
Point(15, 104);
            this.label12.Name = "label12";
            this.label12.Size = new System.Drawing.Size(113,
13);
            this.label12.TabIndex = 36;
            this.label12.Text = "bytes was generated:";
            //
            // Form1
            //
            this.AutoScaleDimensions = new System.Drawing.
SizeF(6F, 13F);
            this.AutoScaleMode = System.Windows.Forms.
AutoScaleMode.Font;
            this.AutoSize = true;
            this.AutoSizeMode = System.Windows.Forms.
AutoSizeMode.GrowAndShrink;
            this.ClientSize = new System.Drawing.Size(624,
596);
            this.Controls.Add(this.panel2);
            this.Controls.Add(this.tabControl1);
            this.Controls.Add(this.button3);
            this.Controls.Add(this.label1);
            this.Controls.Add(this.panel1);
            this.Controls.Add(this.menuStrip1);
            this.Icon = ((System.Drawing.Icon)(resources.
GetObject("$this.Icon")));
            this.MinimumSize = new System.Drawing.Size(640,
634);
            this.Name = "Form1";
            this.Text = "HexRND3";
            this.Load += new System.EventHandler(this.Form1_
Load);
            this.groupBox4.ResumeLayout(false);
            this.groupBox4.PerformLayout();
            this.groupBox1.ResumeLayout(false);
            this.groupBox1.PerformLayout();
            this.groupBox2.ResumeLayout(false);
            this.groupBox2.PerformLayout();
            this.groupBox5.ResumeLayout(false);
            this.groupBox5.PerformLayout();
            ((System.ComponentModel.ISupportInitialize)(this.
pictureBox2)).EndInit();
```

```
        this.menuStrip1.ResumeLayout(false);
        this.menuStrip1.PerformLayout();
        this.tabPage8.ResumeLayout(false);
        this.tabPage8.PerformLayout();
        ((System.ComponentModel.ISupportInitialize)(this.
pictureBox6)).EndInit();
        ((System.ComponentModel.ISupportInitialize)(this.
pictureBox7)).EndInit();
        this.tabPage9.ResumeLayout(false);
        this.tabPage9.PerformLayout();
        ((System.ComponentModel.ISupportInitialize)(this.
pictureBox8)).EndInit();
        ((System.ComponentModel.ISupportInitialize)(this.
pictureBox9)).EndInit();
        this.tabPage2.ResumeLayout(false);
        this.tabControl2.ResumeLayout(false);
        this.tabPage11.ResumeLayout(false);
        this.tabPage11.PerformLayout();
        ((System.ComponentModel.ISupportInitialize)(this.
pictureBox1)).EndInit();
        this.tabPage6.ResumeLayout(false);
        this.tabPage6.PerformLayout();
        ((System.ComponentModel.ISupportInitialize)(this.
pictureBoxA11)).EndInit();
        ((System.ComponentModel.ISupportInitialize)(this.
pictureBoxA12)).EndInit();
        this.tabPage7.ResumeLayout(false);
        this.tabPage7.PerformLayout();
        ((System.ComponentModel.ISupportInitialize)(this.
pictureBoxA22)).EndInit();
        ((System.ComponentModel.ISupportInitialize)(this.
pictureBoxA21)).EndInit();
        this.tabPage10.ResumeLayout(false);
        this.tabPage10.PerformLayout();
        ((System.ComponentModel.ISupportInitialize)(this.
pictureBoxA32)).EndInit();
        ((System.ComponentModel.ISupportInitialize)(this.
pictureBoxA31)).EndInit();
        this.tabPage1.ResumeLayout(false);
        this.groupBox6.ResumeLayout(false);
        this.groupBox6.PerformLayout();
        this.tabControl1.ResumeLayout(false);
        this.panel2.ResumeLayout(false);
        this.panel2.PerformLayout();
        this.ResumeLayout(false);
        this.PerformLayout();
    }
    #endregion
    private System.Windows.Forms.Button button1;
```

```
        private System.Windows.Forms.GroupBox groupBox4;
        private System.Windows.Forms.Button button2;
        private System.Windows.Forms.Label label21;
        private System.Windows.Forms.Label label19;
        private System.Windows.Forms.ComboBox comboBox5;
        private System.Windows.Forms.ComboBox comboBox4;
        private System.Windows.Forms.ComboBox comboBox3;
        private System.Windows.Forms.Panel panel1;
        private System.ComponentModel.BackgroundWorker
backgroundWorker1;
        private System.Windows.Forms.GroupBox groupBox1;
        private System.Windows.Forms.RadioButton radioButton3;
        private System.Windows.Forms.RadioButton radioButton2;
        private System.Windows.Forms.RadioButton radioButton1;
        private System.Windows.Forms.ToolTip toolTip1;
        private System.Windows.Forms.Label label1;
        private System.Windows.Forms.GroupBox groupBox2;
        private System.Windows.Forms.CheckBox checkBox1;
        private System.Windows.Forms.GroupBox groupBox5;
        private System.Windows.Forms.TextBox textBox15;
        private System.Windows.Forms.RadioButton radioButton6;
        private System.Windows.Forms.TextBox textBox13;
        private System.Windows.Forms.RadioButton radioButton4;
        private System.Windows.Forms.RadioButton radioButton5;
        private System.Windows.Forms.Label label2;
        private System.Windows.Forms.PictureBox pictureBox2;
        private System.Windows.Forms.Label label3;
        private System.Windows.Forms.Label label5;
        private System.Windows.Forms.Label label4;
        private System.Windows.Forms.Label label13;
        private System.Windows.Forms.Label label14;
        private System.Windows.Forms.Label label9;
        private System.Windows.Forms.Label label10;
        private System.Windows.Forms.Label label11;
        private System.Windows.Forms.Label label6;
        private System.Windows.Forms.Label label7;
        private System.Windows.Forms.Label label8;
        private System.Windows.Forms.Label label15;
        private System.Windows.Forms.Label label16;
        private System.Windows.Forms.Label label17;
        private System.Windows.Forms.Label label23;
        private System.Windows.Forms.Label label22;
        private System.Windows.Forms.Label label20;
        private System.Windows.Forms.Label label18;
        private System.Windows.Forms.Timer timer1;
        private System.Windows.Forms.MenuStrip menuStrip1;
        private System.Windows.Forms.ToolStripMenuItem
fileToolStripMenuItem;
```

```
        private System.Windows.Forms.ToolStripMenuItem
saveConditionCAinExternalFileToolStripMenuItem;
        private System.Windows.Forms.ToolStripMenuItem
downloadCAConditionFromanExternalFileToolStripMenuItem;
        private System.Windows.Forms.ToolStripSeparator
toolStripSeparator1;
        private System.Windows.Forms.ToolStripMenuItem
ExitfromProgramToolStripMenuItem;
        private System.Windows.Forms.ColorDialog colorDialog1;
        private System.Windows.Forms.SaveFileDialog
saveFileDialog1;
        private System.Windows.Forms.OpenFileDialog
openFileDialog1;
        private System.Windows.Forms.ToolStripMenuItem
referenceToolStripMenuItem;
        private System.Windows.Forms.ToolStripMenuItem
aboutThePprogramToolStripMenuItem;
        private System.Windows.Forms.Button button3;
        private System.Windows.Forms.TabPage tabPage8;
        private System.Windows.Forms.PictureBox pictureBox6;
        private System.Windows.Forms.PictureBox pictureBox7;
        private System.Windows.Forms.Label label41;
        private System.Windows.Forms.Label label42;
        private System.Windows.Forms.Label label43;
        private System.Windows.Forms.Label label44;
        private System.Windows.Forms.Label label45;
        private System.Windows.Forms.Label label46;
        private System.Windows.Forms.Label label47;
        private System.Windows.Forms.TabPage tabPage9;
        private System.Windows.Forms.PictureBox pictureBox8;
        private System.Windows.Forms.PictureBox pictureBox9;
        private System.Windows.Forms.Label label48;
        private System.Windows.Forms.Label label49;
        private System.Windows.Forms.Label label50;
        private System.Windows.Forms.Label label51;
        private System.Windows.Forms.Label label52;
        private System.Windows.Forms.Label label53;
        private System.Windows.Forms.Label label54;
        private System.Windows.Forms.TabPage tabPage2;
        private System.Windows.Forms.TabControl tabControl2;
        private System.Windows.Forms.TabPage tabPage11;
        private System.Windows.Forms.PictureBox pictureBox1;
        private System.Windows.Forms.Label label62;
        private System.Windows.Forms.Label label63;
        private System.Windows.Forms.Label label64;
        private System.Windows.Forms.Label label65;
        private System.Windows.Forms.Label label66;
        private System.Windows.Forms.Label label67;
        private System.Windows.Forms.Label label68;
        private System.Windows.Forms.TabPage tabPage6;
```

```
        private System.Windows.Forms.PictureBox pictureBoxA12;
        private System.Windows.Forms.Label label30;
        private System.Windows.Forms.Label label27;
        private System.Windows.Forms.Label label31;
        private System.Windows.Forms.Label label29;
        private System.Windows.Forms.Label label32;
        private System.Windows.Forms.Label label28;
        private System.Windows.Forms.Label label33;
        private System.Windows.Forms.TabPage tabPage7;
        private System.Windows.Forms.PictureBox pictureBoxA22;
        private System.Windows.Forms.PictureBox pictureBoxA21;
        private System.Windows.Forms.Label label34;
        private System.Windows.Forms.Label label35;
        private System.Windows.Forms.Label label36;
        private System.Windows.Forms.Label label37;
        private System.Windows.Forms.Label label38;
        private System.Windows.Forms.Label label39;
        private System.Windows.Forms.Label label40;
        private System.Windows.Forms.TabPage tabPage10;
        private System.Windows.Forms.PictureBox pictureBoxA32;
        private System.Windows.Forms.PictureBox pictureBoxA31;
        private System.Windows.Forms.Label label55;
        private System.Windows.Forms.Label label56;
        private System.Windows.Forms.Label label57;
        private System.Windows.Forms.Label label58;
        private System.Windows.Forms.Label label59;
        private System.Windows.Forms.Label label60;
        private System.Windows.Forms.Label label61;
        private System.Windows.Forms.TabPage tabPage1;
        private System.Windows.Forms.GroupBox groupBox6;
        private System.Windows.Forms.TabControl tabControl1;
        private System.Windows.Forms.PictureBox pictureBoxA11;
        private System.Windows.Forms.ProgressBar progressBar1;
        private System.Windows.Forms.Button button4;
        private System.Windows.Forms.Timer timer2;
        private System.Windows.Forms.Panel panel2;
        private System.Windows.Forms.Label label26;
        private System.Windows.Forms.Label label25;
        private System.Windows.Forms.Label label24;
        private System.Windows.Forms.Label label12;
        private System.Windows.Forms.Label label172;
        private System.Windows.Forms.Label label174;
        private System.Windows.Forms.Label label169;
        private System.Windows.Forms.Label label170;
        private System.Windows.Forms.Label label171;
        private System.Windows.Forms.Label label175;
        private System.Windows.Forms.Label label173;
        private System.Windows.Forms.CheckBox checkBox2;
    }
}
```

Chapter 7

Processing and Recognition of Images Based on Asynchronous Cellular Automata

ABSTRACT

This chapter discusses the use of asynchronous cellular automata with controlled movement of active cells for image processing and recognition. A time-pulsed image description method is described. Various models and structures of cellular automata for transmitting active signals are presented. The image of the figure is binarized and an active signal moves along its edges. At every moment in time, the active cell of an asynchronous cellular automaton generates a pulse signal. The shape of the generated pulse sequence describes the geometric shape of a flat figure. Methods for describing images of individual plane figures, as well as a method for describing images consisting of many separate geometric objects, are proposed. Cellular automaton is considered as an analogue of the retina of the human visual canal. The circuitry structures of cells of such asynchronous cellular automata are presented, and the software implementation of the proposed methods is also performed. Methods allow one to classify individual geometric image objects.

DOI: 10.4018/978-1-7998-2649-1.ch007

INTRODUCTION

In previous chapters, ACA models were considered in which active signals were transmitted to neighboring cells that are in a given information state, and their neighborhood cells also have a given logical state. Such ACAs have been classified as ACAs with controlled transmission of active signal (Bilan, 2017). In them, the active signal can be transmitted only in the right direction. ACA with this structure can be effectively used in tracking systems. They are also effectively used for image processing and recognition. (Belan, 2011; Belan, & Belan, 2012; Belan, & Belan, 2013; Bilan, 2014).

In addition, ACA with controlled transmission of active signals are effectively used to simulate various processes. Such ACAs are little studied today in terms of various evolutionary paradigms, as well as in terms of their application in scientific research and industry. This chapter discusses the use of ACA with controlled movement of active signals for processing and recognition of 2D images.

The purpose of this chapter is to build methods and means for describing and classifying images of flat geometric shapes based on an ACA with controlled movement of the active signal, which allowed us to create a system that simulates the functions of an analog of the human visual channel.

RETINA ANALOG OF OPTIC CHANNEL
BASED ON CELLULAR AUTOMATA

Modern studies of the physiological structure of the retina of the human visual canal have made it possible to determine the main structural elements and the relationships between them (Schachat, et al 2017; Fineman, 2018;

Figure 1. Simplified horizontal retinal cut model

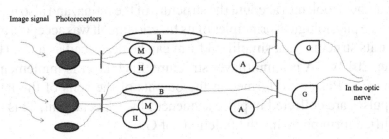

Dowling, 2012). Structurally, the retina can be represented as a horizontal slice (Figure 1).

The first layer of the retina consists of photoreceptors that respond to electromagnetic radiation in the visible part of the spectrum and convert it into electrical (chemical) signals coming to the following functional layers of the retina. Photoreceptors carry out primary processing of optical signals. The subsequent layers of the retina as the main discrete elements contain (Figure 1):

H – horizontal cells;
B – bipolar cells;
M – Mueller cells;
A – amacrine cells;
G – ganglion cells.

All these cells form layers and are interconnected in a certain way. In fact, the retina is a discrete multilayer structure that consists of several groups of homogeneous cells. These cells are intertwined with nerve fibers with many synaptic contacts. As a result of the passage of signals through layers of interwoven cells, and already with semantic content, they enter the ganglion layer. Ganglion cells carry out the collective function of signals from cells of previous layers. At the outputs of the ganglion cells, sequences of impulses are formed (Figure 2), which enter the optic nerve through the optic nerve to the cerebral cortex (Schachat, et al 2017; Fineman, 2018; Dowling, 2012).

At present, the authors are not aware of the results of studies that are devoted to a detailed study of the structure and forms of pulse sequences. The exact relationship between the optical signals at the photoreceptors and the pulse sequences at the outputs of G is not described.

According to the research results, the fact that the retina is a discrete medium that can be represented by a CA based on structural model is obvious. Such a model in rough form was presented in the works (Belan, 2011; Belan, & Belan, 2012) and is multi-layered (Figure 3).

The authors took into account the structure of the retina, and also took into account the physiological principles of its functioning. It was accepted that the retina in its structure in a simplified form can be described as a CA (Belan, & Belan, 2012). CA is a multilayer structure. Each layer is implemented as a simple CA. Pulses are formed at the outputs of the cells of the last CA. These pulses are collected in pulse sequences by a special unit. This unit in a simplified form performs the functions of G.

Figure 2. Forms of electrical signals at the outputs of ganglion cells

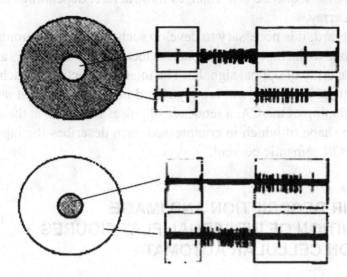

Based on this model, the structure of the retina can be described. However, it does not provide an analogue of the processes of the functioning of the retina. To do this, it is necessary to build the necessary models for each element (cell) of the CA. It is necessary to take into account the interconnected generalized image processing mechanism.

If to consider the retina model as a black box, at the inputs of which there is an array of optical signals, and pulse sequences are formed at the outputs, then the whole question is in the form of a pulse sequence that describes the input array of optical signals. Since the input array can change its parameters,

Figure 3. Retina model based on CA

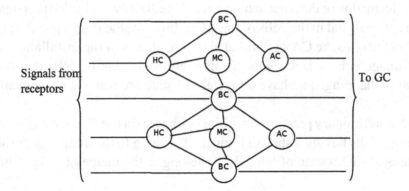

the output pulse sequence also changes its structure, depending on changes in the input array.

In this regard, it is necessary to develop such a cellular environment that would respond to the input array of photodetectors, which receives an optical image (input array of optical signals). The internal structure of each cell and the connections between them are organized in such a way that as a result of the functioning of the CA, a sequence of pulses is formed at the output of the CA, the shape of which in compressed form describes the input image without loss of semantic content.

CONTOUR DESCRIPTION AND IMAGE RECOGNITION OF INDIVIDUAL FLAT FIGURES BASED ON CELLULAR AUTOMATA

It is impossible to describe the image only using the CA. However, a two-dimensional cellular automaton is the main element in the system of description and recognition of images. In an artificial model of the optic canal, the CA can be used as a model of the retina of the eye. Accordingly, the CA must perform the functions of the retina.

To build an artificial model of the retina, an asynchronous cellular automaton with controlled movement of active cells is used. Active cells in such an ACA move only along those cells that have predetermined properties, as well as the desired state of the neighborhood cells. For example, a cell goes into an active state at the next time step if it has a logical "1" state and among the cells in the neighborhood of a given cell there are three cells that have a logical "1" state.

ACA in such a system performs the function of generating a pulse sequence, which in a certain form describes the entire image.

The formation of the pulse sequence is carried out by controlled propagation of the active signal in the cells of the ACA. Initially, the image of a flat figure is projected onto the CA in parallel along the inputs of the installation of all cells. Image is binarized. Noise is also removed. That is, cells that do not belong to the image and have a logical "1" state are reset (set to logical "0" state).

After preliminary preparation, a binary image of a flat figure remains in the CA field. Cells having a state of logical "0" belong to the image background, and cells having a state of logical "1" belong to the image of a flat figure.

The image of a flat figure is a polygon. A polygon is the area described by the outline. The contour contains sides (line segments) connected to the ends of adjacent sides and forming vertices. In fact, the outline describes the semantic content of the polygon. The vertex and two adjacent sides form an angle of a certain size. If the angle at the vertex is less than 90^0, then the vertex is considered convex, and if the angle at the vertex is greater than 90^0, then the vertex is defined as concave.

If you describe the image of a flat figure using the values of the sides, angles and the number of vertices (sides), then you can accurately classify the image as a polygon. The image of polygons can be classified by the number of vertices (sides), a sequence of side lengths and a sequence of angle values at each vertex. The process of classifying images of plane figures as polygons can be represented by the following sequence of steps.

1. Determine the number of steps.
2. The number of concave and convex vertices is determined.
3. The angles at the vertices are determined.
4. The value of the lengths of sides are determine.

Classes are divided by the number of vertices. Each formed class is divided by the values of the angles at the vertices. The resulting subclass is divided by the lengths of the sides at the vertices. For example, a quadrangle may have angles of 90^0 at each vertex. Moreover, it can be divided into squares and rectangles large and small.

An ACA with controlled transmission of active signals is used to describe the contour of the image of a plane figure. Moreover, during the passage of the active signal through the cells belonging to the image contour of a flat figure, pulse signals are generated at the output of each cell. The amplitude of the generated pulse depends on the angle of the active vertex of the image of the figure.

At the initial time, one of the cells of the ACA is selected, which is installed in an active state. As a rule, the extreme cell that belongs to the background is selected. With subsequent time steps, the active signal is transmitted to neighboring cells belonging to the background. With each subsequent time step, the number of active cells belonging to the background increases. During the propagation of active signals through the background cells, pulses at the outputs of the ACA are not formed.

At a certain time step, the active signal reaches the cell that belongs to the edge of the image of a flat figure (contour). As a rule, the nearest cell is the cell belonging to the top of the figure. As soon as the active signal reaches the cell contour of the image, the active background cells are nullified (become inactive). Only one contour cell remains an active cell.

At the following time steps, the active signal propagates only along the cells belonging to the cells of the circuit. This uses the propagation of the active signal in one of the directions of the circuit.

There are two possible ways to propagate an active signal.

- the signal propagates through the extreme cells of the filled image.
- the signal propagates through the selected extreme cells that form the image edge.

In the case of the first mode of active signal propagation, the following mathematical model is used

$$S_{act}(t+1) = \begin{cases} 1, if \vee_{i=1}^{n} X_{act,i}(t) = 1 \, and \, S(t) = 1 \, and \, X_i(t) = 0 \, and \, S_{act}(t) = 0 \\ 0, if \vee_{i=1}^{n} X_{act,i}(t) = 0 \, or \, S_{act}(t) = S_{act}(t-1) = 1 \end{cases},$$

$$(27)$$

where n – number of cells in the neighborhood.

Model (7.1) does not take into account the direction of movement of the active signal at the initial moment of time. The active signal can propagate in two directions from the initial active cell of the circuit. However, this complicates the processing scheme of the generated two pulse sequences. Therefore, an additional circuit or algorithm is used to select the direction of the circuit path by the active signal. An example would be the choice of direction at the initial time by numbering cells in the neighborhood of active cells. For example, the next active cell becomes a neighborhood cell with a lower numbering index (Figure 4).

The model uses the memorization of the active state at the previous two time steps. This is used so that the active signal does not return in the opposite direction of edge bypass. In each active cell of the edge, a pulse signal is formed, the amplitude of which depends on the direction of movement of the active signal. An example of the distribution of directions on Figure 5 is presented.

This distribution is used for both active signal transmission options. An example of the transmission of active signals along the image circuit according to the first embodiment on Figure 6 is shown.

Figure 4. An example of the transmission of an active signal between cells of the circuit at the initial time

Current time step

Next time step

0	0	1
0	Cact 1	1
0	1	1

0	0	Cact 1
0	1	1
0	1	1

Figure 5. An example of the distribution of the directions of movement of the active signal

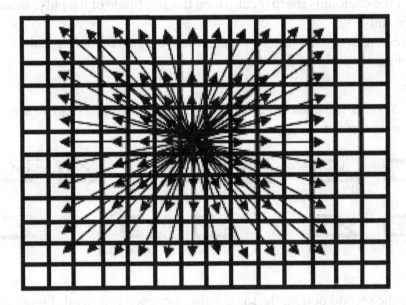

At each time step, the generated pulse signal is presented, and the pulse

Figure 6. Propagation of an active signal through contour cells

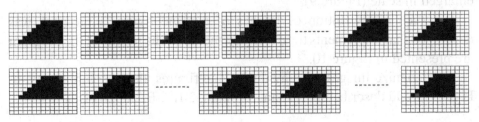

sequence as a whole in Figure 7.

Figure 7. The pulse sequence image for the example shown in Figure 6

As can be seen from Figure 7, pulses with a unit amplitude are formed at each time step since there are no significant kinks in the directions. If the kink of the contour is sharp (peak), then the amplitude of the pulse increases significantly.

An example of the transmission of an active signal according to the second embodiment on Figure 8 is shown.

Figure 8. The distribution of the active signal in the cells of the circuit according to the second option

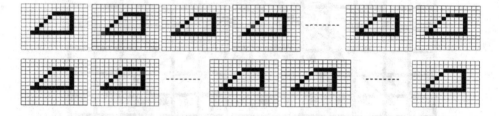

For the second option, the same pulse sequence is formed (Figure 8).

The initially generated impulse sequence contains impulses that are formed at the vertices of the figures and impulses obtained as a result of aliasing. The use of threshold processing in amplitude allows one to obtain a sequence in which there are only pulses that indicate the corresponding vertex and are enlarged in scale (Figure 9).

For such a description of images of a plane figure, a program was developed that implements the described method. The results of the program are presented in Figure 10.

To recognize images of individual plane figures, the structure shown in Figure 11 and described in (Belan, & Belan, 2012; Bilan, 2014).

Figure 9. An example of a generated pulse sequence (a is the initial pulse sequence, b is the pulse sequence after threshold processing)

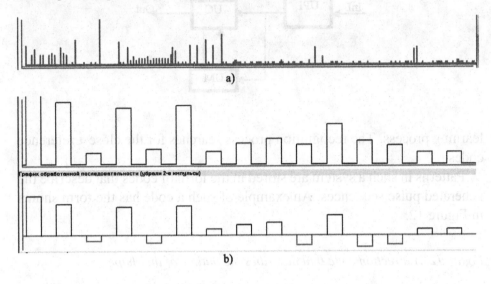

Figure 10. The results of the program simulating the time - pulse image description algorithm

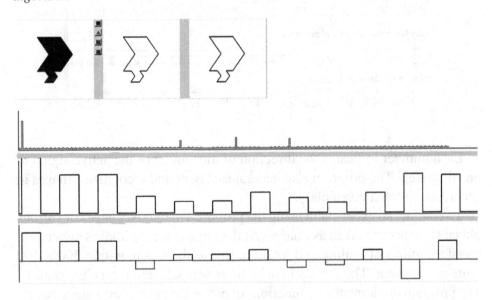

The system learns by forming new classes. New classes are formed with the arrival of a new unknown image. The reference database is formed in the

Figure 11. General image recognition structure

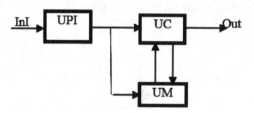

learning process. The recognition process searches for the closest reference codes.

Patterns in such a system are stored in the form of codes that describe the generated pulse sequences. An example of such a code has the form shown in Figure 12.

Figure 12. A direction code that describes the outline of the shape

Triangle codes

Bilan contour detection

> 33224233333322222233333222222333332222233333

von Neumann contour detection

> 33224233333322222233333222222333332222233333

Moore contour detection

> 33344444444333222222333333222222333333222222233

Each number indicates the direction of movement of the active signal in one time step. The program also generates classes and recognizes images in accordance with a given class.

To conduct an experiment using the proposed methods, images of various planar figures were taken as images. At the same time, the images underwent transformations of scaling and rotation. Image sizes were within 300 × 300 units of the raster. The threshold brightness was selected for a clear outline. The program implements the functions of processing pulse sequences, taking into account the removal of pulses with a unit amplitude and pulses with an amplitude of two units. An adapted selection of the amplitude threshold is carried out, which determines the highest recognition percentage. Also, filtering was carried out by the average value of the pulse amplitudes.

Figure 13. An example of subclassing for hexagons

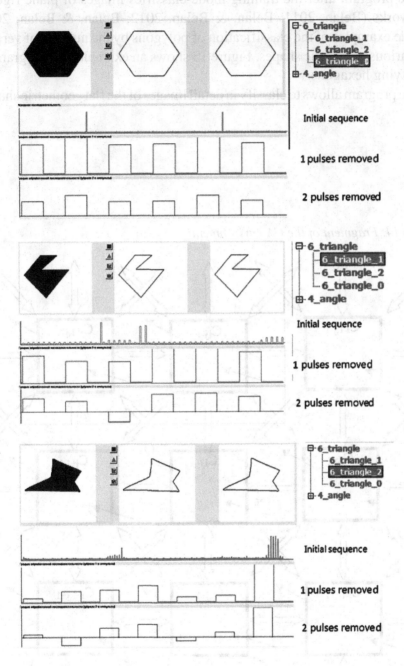

The program after the training mode classifies images of plane figures. The works (Belan, 2011; Belan, & Belan, 2012; Belan, & Belan, 2013) provide examples of the classification of polygons by the number of vertices and various geometric shapes. Figure 13 shows an example of a program for classifying hexagons.

The program allows to classify in detail images of the flat geometric shapes.

Figure 14. Fragment of the CA environment

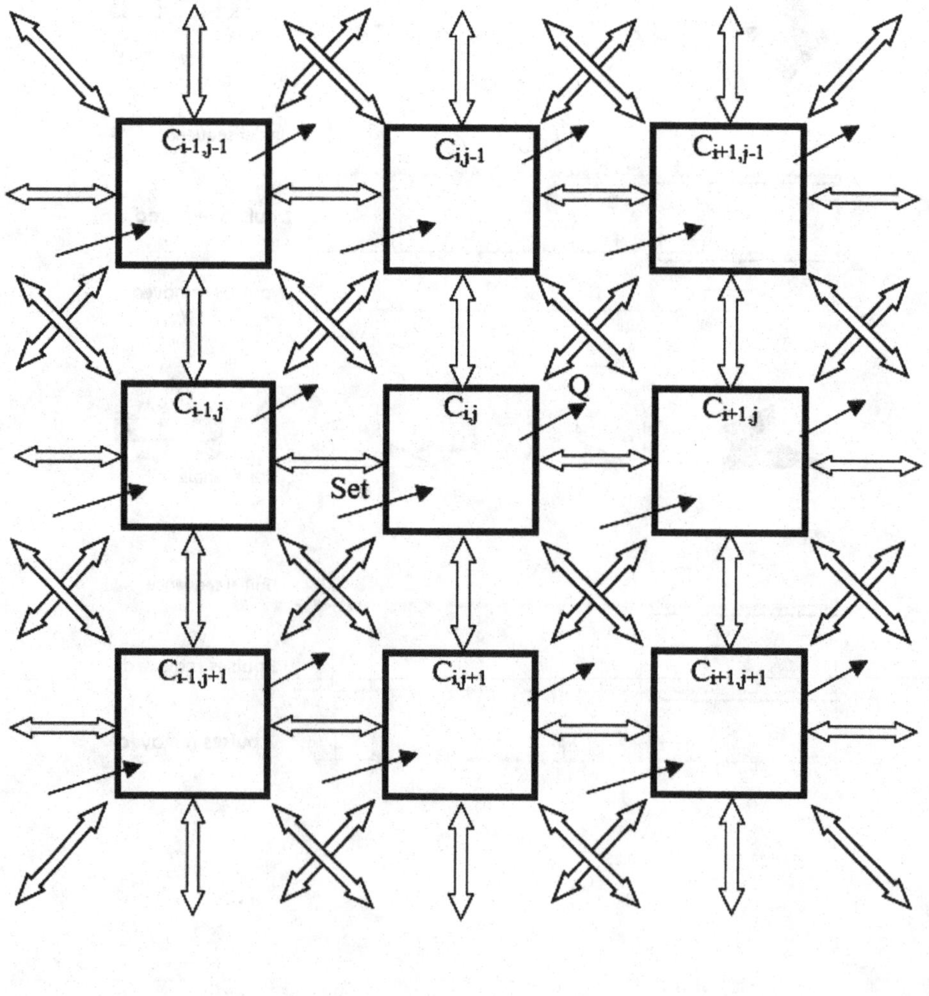

HARDWARE IMPLEMENTATION OF AN ACA CELL WITH CONTROLLED TRANSMISSION OF ACTIVE SIGNALS

The CA environment fragment (Figure 14) contains quadrangular mosaic shaped cells (Cs) that have inputs and outputs connected to the nearest adjacent cells horizontally and vertically. In addition, each cell has an information input (Set) and an output (Q).

According to the schematic illustration (Figure 15), the cell contains information inputs (Set) that connect to the outputs (Q) of the corresponding cells of the neighborhood $C_{i-1,j}$, $C_{i+1,j}$, $C_{i+1,j-1}$, $C_{i-1,j-1}$, $C_{i+1,j+1}$, $C_{i-1,j+1}$, $C_{i,j-1}$, $C_{i1,j+1}$, zeroing input(ZI) of a background cells, active signal inputs (ASI) of active

Figure 15. Schematic representation of a single CA cell

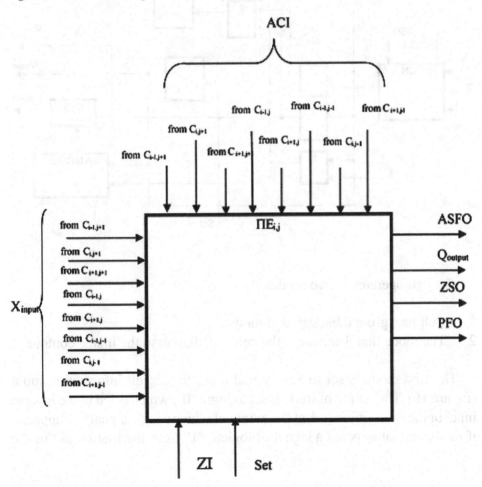

signals, which are connected to active signal formation outputs (ASFO) of neighborhood cells $C_{i-1,j}$, $C_{i+1,j}$, $C_{i+1,j-1}$, $C_{i-1,j-1}$, $C_{i+1,j+1}$, $C_{i-1,j+1}$, $C_{i,j-1}$, $C_{i1,j+1}$. In addition, each cell C environment of CA has zeroing signal output (ZSO), which nullifies the background cells as well pulse formation output (PFO), which is connected to the corresponding input of the two-dimensional OR gate.

The CA cell (Figure 16) contains two analysis units of cell neighborhood (AUCN), active signal forming unit (ASFU), state image memory element (ME), output impulse unit (OIU), and four AND gate.

Figure 16. Functional diagram of the CA cell

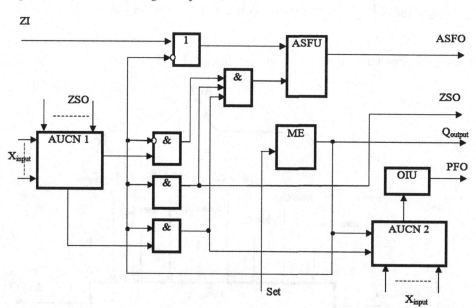

The cell operates in two modes.

1. Cell background background mode.
2. The mode that determines the cell's affiliation to the image contour.

The first mode is set to zero signal input to the cell information input (Figure 16). The output of the C is set to logic "0", which is fed to the inverse input of the first conjunctor, at the output of which a logic signal "1" appears, if its second input is fed a signal of logical "1" from the first output of the first AUCN. A ones signal at the first output of the first AUCN is generated when a logic "1" (excitation signal) is present at one of its control inputs.

A signal of logica "1" from the output of the first conjunctor, via a OR gate, is fed to the first ASFU input, which generates a single signal at the output (excitation signal). So the cell of the background goes into the active state and generates an active signal at the output. Such a cell will be in the active state until a logic signal "1" appears on the input, which is fed to the second ASFU input and returns it to zero. A zero signal is generated at the cell output.

The second cell state mode (image cell membership mode) is set by a single signal from the information input. ME is set to the state of logic "1", which is formed at its output and fed to the first input of the second AND gate. If there is a single signal at the first output of the first AUCN that is fed to the second input of the second AND gate, then a logic "1" is generated at the output of the second AUCN 2. From the output of the second AND gate, a logic signal of "1" is fed to the null output for cells of the background, which is also fed to the null inputs of neighboring cells belonging to the neighborhood. At the first output of the first AUCN, a single signal is generated when at least one of its control inputs has a logical "1" signal coming from the outputs of neighborhood cells. The appearance of a single signal at the output of one of the cells belonging to the image nullifies the active cells belonging to the background.

If the active signal is fed to the control inputs of the first AUCN 1 from cells belonging to the image circuit, then a logical "1" signal is generated at its second output, which is fed to the second input of the third AND gate. The output of the third AND gate generates a logic unit signal that is fed to the second control input of the second AUCN 2, the first control input of which also contains a logic signal "1" from the output ME.

The presence of two units on the control inputs of the second AUCN 2 allows him to analyze the state of neighborhood cells belonging to the contour. Since there are two contour cells belonging to the image contour in the neighborhood, a logical "1" signal at the output of the second AUCN 2 is formed when the adjacent contour cells are not arranged in one line (horizontally, vertically and diagonally). If the cells are located on a diagonal, the output of the second AUCN 2 there is a signal logical "0". If the neighborhood cells that belonging to the contour have an arrangement that forms an acute angle, then a pulse with a higher amplitude than a single one is formed.

The cell functioning algorithm has the following sequence of steps.

1. Determination of a cell belonging is an image or a background.
   1.1. Cell belongs to background. Go to step 2.
   1.2. The cell belongs to the edge of the image. Go to step 7.
2. Determination of the presence of the active signal from the neighborhood cells.
   2.1. Activation signal is not received. Go to step 1.
   2.2. Activation signal was received. Go to step 3.
3. Transition of the cell to the activate state. Formation an active signal at the output. Go to step 4.
4. Determination of the presence of the reset signal.
   4.1. Reset signal is not received. Go to step 4.
   4.2. Reset signal was received. Go to step 5.
5. Cell zeroing. Cell transition to zero state. Go to step 6.
6. The end.
7. Determination of the presence of the active signal.
   7.1. No active signal received. Go to step 7.
   7.2. The active signal was received. Go to step 8.
8. Determination of the cell from which the active signal came.
   8.1. The active signal came from the background cell. Go to step 9.
   8.2. The active signal came from the edge image cell. Go to step 11.
9. Formation of a reset signal for background cells. Go to step 10.
10. Transmission of an active signal to a neighborhood cell wich belonging to the edge and located first clockwise from the cells of the background. Go to step 6.
11. Transmission an active signal to a cell neighborhood that belong to edge and is not active. Go to step 12.
12. Determining the state of the neighborhood.
    12.1. If the contour cells belonging to the edge are located in a straight line, then proceed to point 6.
    12.2. The neighborhood cells belonging to the edge are not placed in a straight line, then proceed to point 13.
13. The formation of a pulse at the output of a cell whose amplitude corresponds to the state of the neighborhood. Go to step 6.

All ACA cells are homogeneous with homogeneous bonds between adjacent cells.

DESCRIPTION AND CLASSIFICATION OF COMPLEX IMAGES BASED ON ACA

Images generally have a complex structure. Complex images consist of many separate objects, which after binarization can be attributed to one of the classes of polygons. To describe a complex image, it is necessary to describe the individual images in it and the relationships between them. To describe a complex image, it is necessary select all the individual objects in the visual scene. After that, a complex image is described and compared with a standard. The structure of the system and the main sequence of actions are described in (Belan, & Belan, 2013).

The following sequence of operations is used to describe and recognize of images.

1. Noise removal and image binarization.
2. Selection objects in an image.
3. Description and classification of each selected object.
4. Determination of the location of selected objects in a complex image and arranging them according to a given parameter.
5. The formation of a common pulse sequence and formed individual sequences.
6. Threshold processing of the primary pulse sequence, as well as averaging and proportionalization of the main quantitative parameters.
7. Recognition of the image by the received pulse sequence.

Figure 17. Example of a general pulse sequence after threshold processing

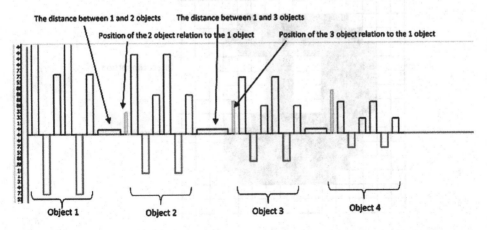

8.   If the image is not recognized, then the formation of a new class.

At the stage of selecting objects, those objects that consist of a small number of cells (less than the threshold) are deleted.

The next step is the formation of a common pulse sequence of the image. This sequence consists of a sequence of pulse sequences of each selected object. The first impulse sequence is an impulse sequence that describes an image object consisting of the largest number of cells. The second pulse sequence describes an object consisting of fewer cells than the previous object, etc. An example of the total pulse sequence on Figure 17 is shown.

In the presented example, the pulse sequences of each object are selected, and the distances between the objects are determined.

Object Selection Based On ACA

Each object in the image is determined by a set of cells with specified physical properties. These cells are combined into one object. This means that each cell of an object has at least one of its cells in a neighborhood. Between all cells there are direct and indirect connections. Direct connections are determined by connections with the cells of the neighborhood, and indirect connections are determined by the fact that signals are transmitted through cells located between two cells of the object. An example of the image of the selected object on Figure 18 is shown.

Figure 18. Example of image of a selected object

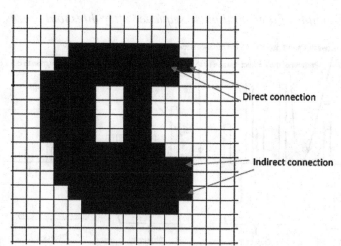

If a cell transmits an active signal to a remote cell of the same object, then all the cells through which the active signal has passed become connected to one object. Moreover, different image objects may consist of cells with the same properties (for example, a binary image).

At the same time, cells of different objects cannot exchange signals, since they are separated by background cells. Another approach described is to determine the cells of the edges of the image of the object and fill it from the edges (Figure 19).

Figure 19. An example of selecting an object by filling it with cells inside the path

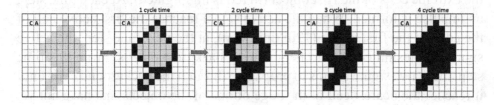

The edge cells by transmitting active signals fill the image of the object from the inside.

Figure 20. An example of selecting objects in a visual scene

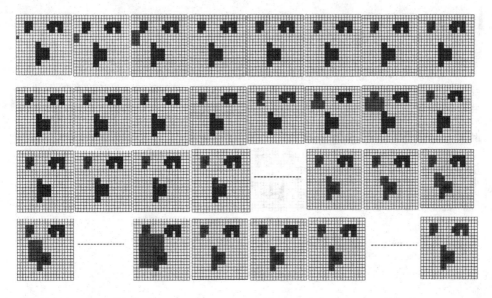

Select Objects in a Complex Image

In a complex image, the sequence of selection of objects does not depend on their size, but depends on their location to the original active cell. Figure 20 shows an example of the selection of objects in the visual scene based on ACA.

As can be seen from the figure, the first active background cell was initially selected and, at subsequent time steps, the active signal propagated through the background cells (in blue is highlighted). At the fourth time step, the active signal reaches the nearest cell in the circuit, which is set to the active state. At this time step, a pulse signal is generated at the output of the ACA. In the next eight time steps, the active signal propagates only to the extreme cells of the first object (in red is highlighted). At each time step, an impulse is formed. When the active signal reaches the first active cell of the edge of the first image (12th time step), at the next time step, the active signal is transmitted only to neighboring background cells. From this time step, the active signal begins to propagate through the background cells until it reaches the nearest cell of the edge of one of the objects that have not yet been selected (15th time step).

Starting from the 16th time step, the active signal begins to propagate only along the extreme cells of the second image object. An impulse sequence is generated that describes the second selected object. After the description of

Figure 21. An example of selecting objects during their subsequent removal

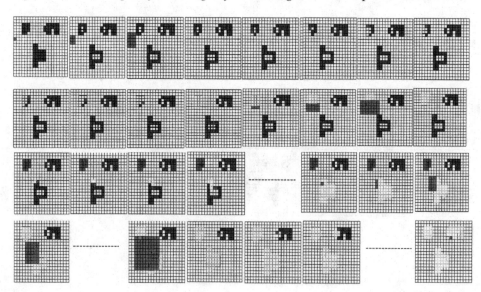

the second object, the third object is selected and the third pulse sequence is formed.

Bidirectional propagation of the active signal through the edge cells of each object is also possible, which reduces processing time and complicates the generated bit sequences.

Another option for selecting objects in the visual scene is when, in the process of moving the active signal, the contour cells that were active in the previous time step are removed. An example of such a selection option on Figure 21 is shown.

Often this option of selecting objects is preferable.

Recognition of Complex Images

When forming a common pulse sequence, the size of the object and the distance between the objects are taken into account. The distance is determined by the propagation time of the active signal through the background cells from object to object. A program is developed that implements this method. The program takes into account the distance between the first object of the visual scene. Each object is described by numbers near the image of each object (Figure 22), which indicate the number of cells, distance and location.

Figure 22. An example of the raw image preprocessing data

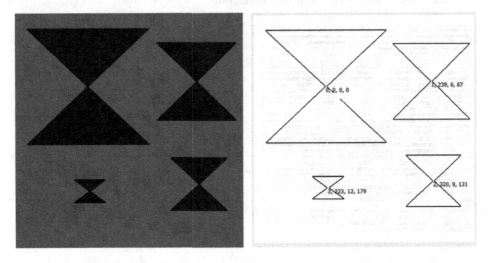

Based on these data, a general sequence is formed.

The system implements two modes: learning mode and recognition mode. Initially, a training set of reference codes is created, which are entered into the memory of the system's standards. At this stage, classes are formed.

In real biological systems, the learning and recognition modes are not separated. In the process of life, the recognition system recognizes those images with which it is learned. If an image arrives at the system input that does not belong to any created class, then a decision is made to form a new class. Known input image makes class more resilient.

In the work, the two modes are not separated. The formation of the class and standards within the class is carried out according to the following characteristics.

1. The number of objects in the visual scene.
2. Object values.
3. Distance between objects.
4. Location of objects.
5. Relations between the selected quantities.

As a result of processing a complex image, a code of directions of movement of the active signal is generated (Figure 23).

Figure 23. The example of complex image code

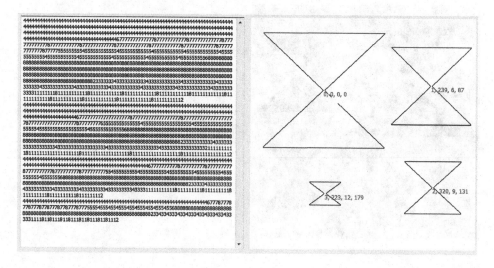

Changing the main parameters leads to the formation of new classes. Classes in the work are divided by the following parameters.

1. The number of individual objects in the visual scene.
2. The size of the object, which is determined by the number of cells that make up the object.
3. Distances between objects, which are determined by the number of cells through which the active signal passes from the object to the object. Least cells are selected.
4. Location of objects.
5. Defined relationships between objects.

Based on these quantitative characteristics, a code is generated that indicates the ratio of the input image to the class.

Classes and subclasses are created in the proposed software in learning mode. In recognition mode, pulse sequences are generated that undergo threshold processing. The time and amplitude characteristics of the processed pulse sequence determine the class to which the input image belongs. An example of the formation of a class tree on Figure 24 is shown.

Figure 24. The example of simple image classification

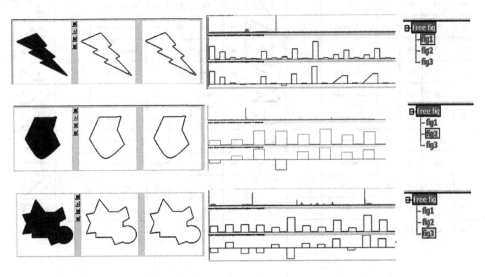

Figure 25. Determination of angles

N	designation	α°	β°	direction	note	figure	N	designation	α°	β°	direction	note	figure
1	1-2	135	225	clockwise			29	5-1	0	360	back		
2	1-3	90	270	clockwise			30	5-2	315	45	counterclock-wise		
3	1-4	45	315	clockwise			31	5-3	270	90	counterclock-wise		
4	1-5	0	360	back			32	5-4	225	135	counterclock-wise		
5	1-6	315	45	counterclock-wise			33	5-6	135	225	clockwise		
6	1-7	270	90	counterclock-wise			34	5-7	90	270	clockwise		
7	1-8	215	145	counterclock-wise			35	5-8	45	315	clockwise		
8	2-1	225	135	counterclock-wise			36	6-1	45	315	clockwise		
9	2-3	135	225	clockwise			37	6-2	0	360	back		
10	2-4	90	270	clockwise			38	6-3	315	45	counterclock-wise		
11	2-5	45	315	clockwise			39	6-4	270	90	counterclock-wise		
12	2-6	0	360	back			40	6-5	225	135	counterclock-wise		
13	2-7	315	45	counterclock-wise			41	6-7	135	225	clockwise		
14	2-8	270	90	counterclock-wise			42	6-8	90	270	clockwise		
15	3-1	270	90	counterclock-wise			43	7-1	90	270	clockwise		
16	3-2	225	135	counterclock-wise			44	7-2	45	315	clockwise		
17	3-4	135	225	clockwise			45	7-3	0	360	back		
18	3-5	90	270	clockwise			46	7-4	315	45	counterclock-wise		
19	3-6	45	315	clockwise			47	7-5	270	90	counterclock-wise		
20	3-7	45	315	back			48	7-6	225	135	counterclock-wise		
21	3-8	315	45	counterclock-wise			49	7-8	135	225	clockwise		
22	4-1	315	45	counterclock-wise			50	8-1	135	225	clockwise		
23	4-2	270	90	counterclock-wise			51	8-2	90	270	clockwise		
24	4-3	225	135	counterclock-wise			52	8-3	45	315	clockwise		
25	4-5	135	225	clockwise			53	8-4	0	360	back		
26	4-6	90	270	clockwise			54	8-5	315	45	counterclock-wise		
27	4-7	45	315	clockwise			55	8-6	270	90	counterclock-wise		
28	4-8	0	360	back			56	8-7	225	135	counterclock-wise		

An example of the formation of subclasses in Figure 13 is presented. All quantitative characteristics are recorded in a table where they can be evaluated at any time.

To implement the recognition method, the raspozn.exe program was created in Delphi (Apendix 2). The program implements a number of procedures, which are described in detail in Apendix 2.

For plotting, the ParNap procedure is used, which determines the direction parameters from the sequence of directions. This procedure uses angle templates from the file "corners.txt", an illustration of which is the file "corners. xls". The screenshot of the file has the name "corners.png" (Figure 25)

CONCLUSION

The use of ACA with controlled transmission of active signals allowed us to create a method for the selection and description of objects in the visual scene. Based on this method, a system for recognizing and describing complex images has been developed. The method is most effective for simple images, and is also acceptable, since it can be used for a wide range of classes.

REFERENCES

Belan, S., & Belan, N. (2012). Use of Cellular Automata to Create an Artificial System of Image Classification and Recognition. *LNCS, 7495*, 483–493.

Belan, S., & Belan, N. (2013). Temporal-Impulse Description of Complex Image Based on Cellular Automata. In PaCT2013 (LNCS, Vol. 7979, pp. 291-295). Springer-Verlag.

Belan, S. N. (2011). Specialized cellular structures for image contour analysis. *Cybernetics and Systems Analysis, 47*(5), 695–704. doi:10.100710559-011-9349-8

Bilan, S. (2014). Models and hardware implementation of methods of Pre-processing Images based on the Cellular Automata. *Advances in Image and Video Processing, 2*(5), 76–90. doi:10.14738/aivp.25.561

Bilan, S. (2017). *Formation Methods, Models, and Hardware Implementation of Pseudorandom Number Generators: Emerging Research and Opportunities.* IGI Global.

Dowling, J. E. (2012). The Retina: An Approachable Part of the Brain (2nd ed.). Belknap Press.

Fineman, M. (2018). *Retina (Color Atlas and Synopsis of Clinical Ophthalmology)* (3rd ed.). LWW.

Schachat, P. A., Wilkinson, C. P., Hinton, D. R., Wiedemann, P., Freund, K. B., & Sarraf, D. (2017). *Ryan's Retina: 3 Volume Set*. Elsevier.

APPENDIX

To implement the recognition method, the raspozn.exe program was created in Delphi, described in this appendix. The program implements a number of procedures, which are described in detail here. For plotting, the ParNap procedure is used, which determines the direction parameters from the sequence of directions. This procedure uses angle templates from the file "corners.txt", an illustration of which is the file "corners.xls". 17 fragments of the program are described.

This section describes the functioning of the raspozn.exe program. The first operation that implements the algorithm is to remove noise and normalize the background image. The program starts by performing the "Remove Background" procedure. The listing of this procedure is described below.

```
procedure Tglavn.BitBtn21Click(Sender: TObject);
var proc: Integer;
    limit, Col: TColor;
    Mcol1Fon, MColLimFon: TMasCol;
    i, j: Integer;
begin
On_Off(False);
If Na_BMP = '' Then
 ShowMessage('Image not uploaded!')
 else begin
  proc:= TrackBar1.Position;
  limit:= Round(16777215 * proc / 100);
  SetLength(MColLimFon, n, m);
  SetLength(MFonIsh, n, m);
  Mcol1Fon:= MIsh;
  For i:= 0 to n - 1 do
   For j:= 0 to m - 1 do
    begin
     Col:= MCol1Fon[i, j];
     If Col > limit Then Col:= 16777215;
     MColLimFon[i, j]:= Col;
     Image1.Canvas.Pixels[i, j]:= MColLimFon[i, j];
    end;
    MFonIsh:= MColLimFon;
 On_Off(True);
 Fl_Fon:= True;
```

```
  end;
end;
```

Initially, the image is binarized. Listing of the procedure for converting a color image into a bit (two-color):

```
procedure Tglavn.BitBtn18Click(Sender: TObject);
var proc:Integer;
    limit, Col: TColor;
    Mcol1, MColLim: TMasCol;
    i, j: Integer;
begin
If Na_BMP = '' Then
 ShowMessage('Image not uploaded!')
 else begin
  proc:= TrackBar1.Position;
  limit:= Round(16777215 * proc / 100);
  SetLength(MColLim, n, m);
  SetLength(M1, n, m);
  If Fl_Fon Then Mcol1:= MFonIsh else Mcol1:= MIsh;
  For i:= 0 to n - 1 do
   For j:= 0 to m - 1 do
    begin
     Col:= MCol1[i, j];
     If Col > limit Then Col:= 16777215
                  else Col:= 0;
     MColLim[i, j]:= Col;
     Image1.Canvas.Pixels[i, j]:= MColLim[i, j];
    end;
  For i:= 0 to n - 1 do
   For j:= 0 to m - 1 do
    begin
     Col:= Image1.Canvas.Pixels[i, j];
     If Col = clBlack Then M1[i, j]:= 1 Else M1[i, j]:= 0;
    end;
 On_Off(True);
 Fl_Bit:= True;
 end;
end;
```

In the resulting binary image, the outline is selected. Outline selection procedure:

```
Procedure Tglavn.Kontur(Sender: TObject);
var i, j, k, c: Integer;
begin
ObhodSlevaSverhu(M1);
```

```
For i:= 0 to n - 1 do
 For j:= 0 to m - 1 do
  begin
   Gauge1.Progress:= Gauge1.Progress + 1;
   Application.ProcessMessages;
   If M1[i, j] = 1 Then
    If (ResPapa[i, j][2] = '0') or (ResPapa[i, j][4] = '0')or
       (ResPapa[i, j][6] = '0') or (ResPapa[i, j][8] = '0')
      Then Mc[i, j]:= 1
      else Mc[i, j]:= 0
    else Mc[i, j]:= 0;
end;
end;
```

According to the obtained contours, the signal is transmitted and the direction of movement is determined. The nested procedure ObhodSlevaSverhu has the form:

```
Procedure ObhodSlevaSverhu(M1: Variant);
Var
  i, j: Integer;
  k: Word;
  sb: String[8];
Begin
  For i:= 1 to n - 2 do
   For j:= 1 to m - 2 do
    begin
     If M1[i, j] = 1 Then
      begin
     sb:= '';
     If M1[i - 1, j - 1] = 1 Then sb:= sb + '1'
                   Else sb:= sb + '0';
     If M1[i, j - 1] = 1 Then sb:= sb + '1'
                   Else sb:= sb + '0';
     If M1[i + 1, j - 1] = 1 Then sb:= sb + '1'
                   Else sb:= sb + '0';
     If M1[i + 1, j] = 1 Then sb:= sb + '1'
                   Else sb:= sb + '0';
     If M1[i + 1, j + 1] = 1 Then sb:= sb + '1'
                   Else sb:= sb + '0';
     If M1[i, j + 1] = 1 Then sb:= sb + '1'
                   Else sb:= sb + '0';
     If M1[i - 1, j + 1] = 1 Then sb:= sb + '1'
                   Else sb:= sb + '0';
     If M1[i - 1, j] = 1 Then sb:= sb + '1'
                   Else sb:= sb + '0';
     ResPapa[i, j]:= sb;
     end;
```

223

```
      end;
End;
```

Depending on the setting of the switch, the procedure for selecting the direction method is used:

```
If CheckBox4.Checked Then Napravl_Mish(Image2, Mc, Posl, 2)
else Napravl(Image2, Mc, Posl, 2);
```

Listing of the Napravl_Mish procedure with the "with target" switch installed is of the form:

```
Procedure Napravl_Mish(im: TImage; MM: TmasF; var Posl: String;
Mathod: Byte);
type TGLnap = record
              kol: Integer;
              nap: Char;
            end;
var
 i, j, iNach, jNach, imin, imax, jmin, jmax, i1, j1, it, jt, k,
Nach, Napr: Integer;
 deltai, deltaj, righ, bot: Integer;
 Fl, Fl_1, Fl_2: Boolean;
 fpr: TextFile;
 GLnap, GLi, nmax, Dlina, Dlpr, MMend: Integer;
 MMpr: TmasF;
 Mnapr: array of TGLnap;
 MMnap: array of array of Char;
 Nap, Ch: Char;
 sn: String;
begin
FnMish:= glavn.LabeledEdit3.Text;
Misheny(FnMish);
If not FLmish Then
 begin
  Application.MessageBox('Directions with a target cannot be
formed!',
                        'Missing file!', MB_OK + MB_
ICONINFORMATION);
  Exit;
 end;
Fl:= False;
Dlina:= 0;
For jt:= 0 to n - 1 do
 For it:= 0 to m - 1 do
  If MM[it, jt] = 1 Then Inc(Dlina);
glavn.Gauge1.MinValue:= 0; glavn.Gauge1.MaxValue:= Dlina;
```

```
For i:= 0 to n - 1 do
 begin
  For j:= 0 to m - 1 do
   If MM[i, j] = 1 Then
    begin
     Application.ProcessMessages;
     iNach:= i; jNach:= j;
     Fl:= True;
     imin:= i - (r div 2 - 1);
     jmin:= j - (r div 2);
     Break;
    end;
  If Fl Then Break;
 end;
glavn.Gauge1.Progress:= glavn.Gauge1.Progress + 1;
If imin < 0 Then begin deltai:= Abs(imin); imin:= 0; end else
deltai:= 0;
If jmin < 0 Then begin deltaj:= Abs(jmin); jmin:= 0; end else
deltaj:= 0;
righ:= imin + r;
If righ > n Then deltai:= righ - n
    else righ:= 0;
bot:= jmin + r;
If bot > m Then deltaj:= bot - m
    else bot:= 0;
SetLength(MMpr, r, r);
SetLength(MMnap, r, r);
For j1:= 0 to r - 1 - deltaj do
 For i1:= 0 to r - 1 - deltai do
  MMpr[j1, i1]:= MM[i1 + imin, j1 + jmin];
MM[i, j]:= 3;
SetLength(Mnapr, nna);
For j1:= 0 to r - 1 do
 For i1:= 0 to r - 1 do
  MMnap[j1, i1]:= '0';
For i:= 0 to nna - 1 do Mnapr[i].nap:= IntToStr(i + 1)[1];
For jt:= 0 to r - 1 - deltaj do
 For it:= 0 to r - 1 - deltai do
  begin
  If MMpr[jt, it] = 1 Then
   begin
    Ch:= Mmish[jt + deltaj - bot, it + deltai - righ];
    MMnap[jt, it]:= Ch;
    Case Ch of
      '1': begin Inc(Mnapr[0].kol); end;
      '2': begin Inc(Mnapr[1].kol); end;
      '3': begin Inc(Mnapr[2].kol); end;
      '4': begin Inc(Mnapr[3].kol); end;
      '5': begin Inc(Mnapr[4].kol); end;
```

```
    '6': begin Inc(Mnapr[5].kol); end;
    '7': begin Inc(Mnapr[6].kol); end;
    '8': begin Inc(Mnapr[7].kol); end;
   end;
  end;
 end;
GLnap:= 0;   GLi:= 0;
Posl:= '';
For i:= 0 to nna - 1 do
 If Mnapr[i].kol > GLnap Then begin GLnap:= Mnapr[i].kol; GLi:=
i; end;
Posl:= Mnapr[GLi].nap;
 / / - - - - - - - - - - - - - control BEGINNING - - - - - - - -
- - - - - - - - -
If glavn.CheckBox5.Checked Then
 begin
AssignFile(fpr, CurDir + 'contr.txt');
Rewrite(fpr);
Writeln(fpr, 'Common matrix clipping');
For jt:= 0 to r - 1 - deltaj do
 For it:= 0 to r - 1 - deltai do
  If it = r - 1 - deltai Then Writeln(fpr, MMpr[jt, it]) else
Write(fpr, MMpr[jt, it]);
Writeln(fpr, 'Resulting matrix after comparison');
For jt:= 0 to r - 1 - deltaj do
 For it:= 0 to r - 1 - deltai do
  If it = r - 1 - deltai Then Writeln(fpr, MMnap[jt, it]) else
Write(fpr, MMnap[jt, it]);
Writeln(fpr, 'Target Matrix');
For jt:= 0 to r - 1 - deltaj do
 For it:= 0 to r - 1 - deltai do
  If it = r - 1 - deltai Then Writeln(fpr, Mmish[jt + deltaj,
it + deltai]) else Write(fpr, Mmish[jt + deltaj, it + deltai]);
CloseFile(fpr);
 end; / /  If glavn.CheckBox5.Checked
 / / - - - - - - - - - - - - - - END control - - - - - - - - - - -
- - - - -
i:= iNach; j:= jNach;
Napr:= 0; Nach:= 0;   MMend:= 0;
Fl:= True;   Dlpr:= 0;
repeat
  Inc(Dlpr);
  Application.ProcessMessages;
  Fl_1:= False;
  Case Mathod of
1, 2: begin
    i1:= 0;
    repeat
      Inc(i1);
```

```
Case i1 of
 1: begin Dec(i); Dec(j); end;
 2: begin Dec(j); end;
 3: begin Inc(i); Dec(j); end;
 4: begin Inc(i); end;
 5: begin Inc(i); Inc(j); end;
 6: begin Inc(j); end;
 7: begin Dec(i); Inc(j); end;
 8: begin Dec(i); end;
end;
If MM[i, j] = 3 Then MMend:= MM[i, j];
If (MM[i, j] = 1) or (MMend = 3) Then
 begin
  Fl_1:= True;
  Break;
 end;
Case i1 of
 1: begin Inc(i); Inc(j); end;
 2: begin Inc(j); end;
 3: begin Dec(i); Inc(j); end;
 4: begin Dec(i); end;
 5: begin Dec(i); Dec(j); end;
 6: begin Dec(j); end;
 7: begin Inc(i); Dec(j); end;
 8: begin Inc(i); end;
end;
If i1 = 8 Then
 begin
  j1:= 0; Fl_2:= False;
  MM[i, j]:= 4;
  repeat
   Inc(j1);
   Case j1 of
     1: begin Dec(i); Dec(j); end;
     2: begin Dec(j); end;
     3: begin Inc(i); Dec(j); end;
     4: begin Inc(i); end;
     5: begin Inc(i); Inc(j); end;
     6: begin Inc(j); end;
     7: begin Dec(i); Inc(j); end;
     8: begin Dec(i); end;
   end;
   If MM[i, j] = 3 Then MMend:= MM[i, j];
   If (MM[i, j] = 2)or(MMend = 3) Then
    begin
     Fl_2:= True;
     Break;
    end;
   Case j1 of
```

```
      1: begin Inc(i); Inc(j); end;
      2: begin Inc(j); end;
      3: begin Dec(i); Inc(j); end;
      4: begin Dec(i); end;
      5: begin Dec(i); Dec(j); end;
      6: begin Dec(j); end;
      7: begin Inc(i); Dec(j); end;
      8: begin Inc(i); end;
     end;
    until Fl_2;
    end;
  until Fl_1 or (i1 = 8);
  end;
3: begin
  i1:= 0;
  repeat
   Inc(i1);
   If i1 mod 2  = 0 Then
   begin
   Case i1 of
    2: begin Dec(j); end;
    4: begin Inc(i); end;
    6: begin Inc(j); end;
    8: begin Dec(i); end;
   end;
   If MM[i, j] = 3 Then MMend:= MM[i, j];
   If (MM[i, j] = 1) or (MMend = 3) Then Break;
   Case i1 of
    2: begin Inc(j); end;
    4: begin Dec(i); end;
    6: begin Dec(j); end;
    8: begin Inc(i); end;
   end;
   end;
   If i1 = 8 Then
    begin
     j1:= 0; Fl_2:= False;
     MM[i, j]:= 4;
     repeat
      Inc(j1);
      Case j1 of
       1: begin Dec(i); Dec(j); end;
       2: begin Dec(j); end;
       3: begin Inc(i); Dec(j); end;
       4: begin Inc(i); end;
       5: begin Inc(i); Inc(j); end;
       6: begin Inc(j); end;
       7: begin Dec(i); Inc(j); end;
       8: begin Dec(i); end;
```

```
          end;
          If MM[i, j] = 3 Then MMend:= MM[i, j];
          If (MM[i, j] = 2)or(MMend = 3) Then
           begin
            Fl_2:= True;
            Break;
           end;
          Case j1 of
           1: begin Inc(i); Inc(j); end;
           2: begin Inc(j); end;
           3: begin Dec(i); Inc(j); end;
           4: begin Dec(i); end;
           5: begin Dec(i); Dec(j); end;
           6: begin Dec(j); end;
           7: begin Inc(i); Dec(j); end;
           8: begin Inc(i); end;
          end;
        until Fl_2;
       end;
     until Fl_1 or (i1 = 8);
    end;
  end;
  glavn.Gauge1.Progress:= glavn.Gauge1.Progress + 1;
  imin:= i - (r div 2 - 1);
  jmin:= j - (r div 2);
  If imin < 0 Then begin deltai:= Abs(imin); imin:= 0; end
else deltai:= 0;
  If jmin < 0 Then begin deltaj:= Abs(jmin); jmin:= 0; end
else deltaj:= 0;
  righ:= imin + r - n;
  If righ > 0 Then deltai:= righ
       else righ:= 0;
  bot:= jmin + r - m ;
  If bot > 0 Then deltaj:= bot
       else bot:= 0; end;
  For j1:= 0 to r - 1 - deltaj do
   For i1:= 0 to r - 1 - deltai do
    MMpr[j1, i1]:= MM[i1 + imin, j1 + jmin];
  MM[i, j]:= 2;
  For i1:= 0 to nna - 1 do Mnapr[i1].kol:= 0;
  For jt:= 0 to r - 1 - deltaj do
   For it:= 0 to r - 1 - deltai do
    begin
    MMnap[jt, it]:= '0';
    If MMpr[jt, it] = 1 Then
     begin
      Ch:= Mmish[jt + deltaj - bot, it + deltai - righ];
      MMnap[jt, it]:= Ch;
      Case Ch of
```

```
     '1': begin Inc(Mnapr[0].kol); end;
     '2': begin Inc(Mnapr[1].kol); end;
     '3': begin Inc(Mnapr[2].kol); end;
     '4': begin Inc(Mnapr[3].kol); end;
     '5': begin Inc(Mnapr[4].kol); end;
     '6': begin Inc(Mnapr[5].kol); end;
     '7': begin Inc(Mnapr[6].kol); end;
     '8': begin Inc(Mnapr[7].kol); end;
    end;
   end;
  end;
GLnap:= 0;   GLi:= 0;
For i1:= 0 to nna - 1 do
  If Mnapr[i1].kol > GLnap Then begin GLnap:= Mnapr[i1].kol;
GLi:= i1; end;
  If MMend = 0 Then Posl:= Posl + Mnapr[GLi].nap;
  If Fl Then Fl:= False;
 / / - - - - - - - - - - - - - - control BEGINNING - - - - - - - - -
- - - - - - - - -
If glavn.CheckBox5.Checked Then
 begin
AssignFile(fpr, CurDir + 'contr' + IntToStr(Dlpr) + '.txt');
Rewrite(fpr);
Writeln(fpr, 'Common matrix clipping');
For jt:= 0 to r - 1 - deltaj do
 For it:= 0 to r - 1 - deltai do
  If it = r - 1 - deltai Then Writeln(fpr, MMpr[jt, it]) else
Write(fpr, MMpr[jt, it]);
Writeln(fpr, 'Resulting matrix after comparison');
For jt:= 0 to r - 1 - deltaj do
 For it:= 0 to r - 1 - deltai do
  If it = r - 1 - deltai Then Writeln(fpr, MMnap[jt, it]) else
Write(fpr, MMnap[jt, it]);
Writeln(fpr, 'Target Matrix');
For jt:= 0 to r - 1 - deltaj do
 For it:= 0 to r - 1 - deltai do
  If it = r - 1 - deltai Then Writeln(fpr, Mmish[jt + deltaj,
it + deltai]) else Write(fpr, Mmish[jt + deltaj, it + deltai]);
{For jt:= 0 to n - 1 do
 For it:= 0 to m - 1 do
  If it = r - 1 Then Writeln(fpr, MM[it, jt]) else Write(fpr,
MM[it, jt]); }
CloseFile(fpr);
end;
 / / - - - - - - - - - - - - - END control - - - - - - - - - - - -
- - - - -
until (MMend = 3);
If MMend <> 3 Then ShowMessage('Attention! The beginning of the
contour was not found!' + #13 + 'Exit at the end of the
```

```
length.');
sn:= ChangeFileExt(Na_BMP, '.txt');
AssignFile(fpr, CurDir + 'posl_' + IntToStr(Mathod) + '_' +
sn);
Rewrite(fpr);
Writeln(fpr, Posl);
CloseFile(fpr);
For jt:= 0 to n - 1 do
 For it:= 0 to m - 1 do
  If MM[it, jt] > 0 Then MM[it, jt]:= 1;
end;
```

Bypass is carried out using the formed target. Listing the nested target creation procedure:

```
Procedure Misheny(nfm: String);
var s: String;
    i, j, dm: Integer;
begin
AssignFile(fmish, nfm);
If FindFirst(nfm, faArchive, SR) = 0 Then
 begin
  Reset(fmish);
  dm:= StrToInt(nfm[Length(nfm) - 8] + nfm[Length(nfm) - 7]);
  nna:= StrToInt(nfm[Length(nfm) - 5] + nfm[Length(nfm) - 4]);
  r:= dm;
  SetLength(Mmish, dm, dm);
  For i:= 1 to dm do
   begin
    Readln(fmish, s);
    For j:= 1 to dm do Mmish[i - 1, j - 1]:= s[j];
   end;
  CloseFile(fmish);
  FLmish:= True;
 end
 else
  begin
   s:= 'No file' + nfm + 'with a target!';
   Application.MessageBox(PChar(s), 'Missing file!', MB_OK +
MB_ICONINFORMATION);
  end;
end;
```

Listing of the Napravl procedure with the switch installed looks like:

```
procedure Napravl(im: TImage; MM: Variant; var Posl: String;
Mathod: Byte);
```

231

```
var
 i, j, iNach, jNach, i1, j1, i2, j2, k, Nach, Napr: Integer;
 Fl: Boolean;
begin
 Fl:= False;
 For i:= 0 to n - 1 do
  begin
    For j:= 0 to m - 1 do
     If MM[i, j] = 1 Then begin iNach:= i; jNach:= j; Fl:= True;
Break; end;
    If Fl Then Break;
   end;
 im.Canvas.Brush.Color:= clRed;
 im.Canvas.Rectangle(iNach * 10 + 1, jNach * 10 + 1, iNach * 10
+ 9, jNach * 10 + 9);
 i:= iNach; j:= jNach; i1:= 0; j1:= 0;
 Posl:= '';
 glavn.StaticText6.Visible:= True;
 glavn.Shape1.Visible:= True; glavn.Shape2.Visible:= True;
 glavn.Label4.Visible:= True; glavn.Label5.Visible:= True;
 ObhodSlevaSverhuNapr(MM);
 Napr:= 0; Nach:= 0;
 Fl:= True;
 While (i <> iNach) or (j <> jNach) or Fl do
  begin
   Case Napr of
    1: Nach:= 7;
    2: Nach:= 8;
    3: Nach:= 1;
    4: Nach:= 2;
    5: Nach:= 3;
    6: Nach:= 4;
    7: Nach:= 5;
    8: Nach:= 6;
    0: Nach:= 2;
   End;
   Napr:= PoiskNapr(ResKontur[i, j], Nach, Mathod);
   Posl:= Posl + IntToStr(Napr);
   Case (Napr) of
    1: begin Dec(i); Dec(j); end;
    2: begin Dec(j); end;
    3: begin Inc(i); Dec(j); end;
    4: begin Inc(i); end;
    5: begin Inc(i); Inc(j); end;
    6: begin Inc(j); end;
    7: begin Dec(i); Inc(j); end;
    8: begin Dec(i); end;
   End;
   If Fl Then
```

```
    begin
      Fl:= False;
      im.Canvas.Brush.Color:= clYellow;
      im.Canvas.Rectangle(i * 10 + 1, j * 10 + 1, i * 10 + 9, j
* 10 + 9);
      end;
    end;
end;
```

The embedded procedure ObhodSlevaSverhuNapr has the form:

```
Procedure ObhodSlevaSverhuNapr(M1: Variant);
Var
  i, j: Integer;
  k: Word;
  sb: String[8];
Begin
  For i:= 1 to n - 2 do
   For j:= 1 to m - 2 do
     begin
       If M1[i, j] = 1 Then
        begin
         sb:= '';
         If M1[i - 1, j - 1] = 1 Then sb:= sb + '1'
                     Else sb:= sb + '0';
         If M1[i, j - 1] = 1 Then sb:= sb + '1'
                     Else sb:= sb + '0';
         If M1[i + 1, j - 1] = 1 Then sb:= sb + '1'
                     Else sb:= sb + '0';
         If M1[i + 1, j] = 1 Then sb:= sb + '1'
                     Else sb:= sb + '0';
         If M1[i + 1, j + 1] = 1 Then sb:= sb + '1'
                     Else sb:= sb + '0';
         If M1[i, j + 1] = 1 Then sb:= sb + '1'
                     Else sb:= sb + '0';
         If M1[i - 1, j + 1] = 1 Then sb:= sb + '1'
                     Else sb:= sb + '0';
         If M1[i - 1, j] = 1 Then sb:= sb + '1'
                     Else sb:= sb + '0';
         ResKontur[i, j]:= sb;
       end;
     end;
End;
```

The embedded procedure PoiskNapr has the form:

```
Function PoiskNapr(Okr: String; Nach: Shortint; Method: Byte):
Byte;
Var
  i, j: Integer;
  s: String[8];
Begin
i:= Nach; j:= 0;
 While j <> 8 do
   begin
    If i = 9 Then i:= 1;
    Case Method of
      1: If Okr[i] = '1' Then Break;
      2: If Okr[i] = '1' Then Break;
      3: If i mod 2  = 0 Then If Okr[i] = '1' Then Break;
    end;
    Inc(i); Inc(j);
   end;
 PoiskNapr:= i;
End;
```

Recognition procedure:

```
procedure TRaspozF.FormShow(Sender: TObject);
const sc = 'Outline not recognized';
var Eio, i, j, j1, num: Integer;
    Flras_b, Flras_n, Flras_m, FlPolras_b, FlPolras_n,
FlPolras_m: Boolean;
    KodPosl, Kod_Dict, Kod_Pov, Kod_Sdvig: TKodAr;
    s_Dict, ParMi: String;
begin
Left:= 0; Top:= 0;
Width:= 1024; Height:= 730;
If glavn.CheckBox4.Checked Then
  ParMi:= IntToStr(r) + '_' + IntToStr(nna)
 else
  begin
   glavn.ClearScr(Sender, Image2);
   glavn.ClearScr(Sender, Image1);
   Draw_kontur(Posl);
   DrawF(Image2, MPosl);
  end;
LabeledEdit1.Text:= Posl;
LabeledEdit2.Text:= PoslN;
LabeledEdit3.Text:= PoslM;
Image1.Width:= glavn.Image1.Width;
Image1.Height:= glavn.Image1.Height;
Image2.Width:= glavn.Image1.Width;
Image2.Height:= glavn.Image1.Height;
```

```
ShapkaStringGrid(StringGrid1);
ShapkaStringGrid(StringGrid2);
ShapkaStringGrid(StringGrid3);
glavn.Kod(Posl, KodPosl);
If glavn.CheckBox2.Checked Then
 Ispr_Odin_Osh(KodPosl);
VseVarianty(KodPosl, VseVar);
AssignFile(fb, CurDir + ndict_B);
{$I - }Reset(fb); {$I + }
Eio:= IOResult;
If Eio = 0 Then
 begin
  For i:= 0 to FileSize(fb) - 1 do
   begin
    If FlPolras_b Then Break;
    If not Eof(fb) Then Read(fb, dict_b) else Break;
    If (glavn.CheckBox4.Checked) and (dict_B.prim <> ParMi)
Then Continue;
    s_Dict:= dict_B.napr;
    num:= dict_B.num;
    For j:= 2 to num do
     begin
      Read(fb, dict_b);
      s_Dict:= s_Dict + dict_B.napr;
      end;
    glavn.Kod(s_Dict{dict_B.napr}, Kod_Dict);
    For j:= 0 to Length(VseVar) - 1 do
     begin
      Sravnenie(VseVar[j].MasPoint, Kod_Dict, Flras_b,
FlPolras_b);
      If FlPolras_b Then
       begin
        Vyvod_StringGrid(StringGrid1, s_Dict, dict_B, 0, 0, 0,
VseVar[j].Prim);
        BitBtn8.Visible:= False;
        Break;
       end
      else
        If Flras_b Then
         Vyvod_StringGrid(StringGrid1, s_Dict, dict_B, 0, 0, 0,
VseVar[j].Prim);
     end;
   end;
 end;
Benchmark(LabeledEdit5, dtrasp, True);
end;
```

The listing of the Draw_kontur embedded procedure has the form:

```
procedure Draw_kontur(s: String);
var i, L, tb, lr, x, y: Integer;
    spr: String;
    yma, ymi, xma, xmi: Integer;
begin
L:= Length(s);
tb:= 0; lr:= 0;
For x:= 0 to n - 1 do
 For y:= 0 to m - 1 do
   MPosl[x, y]:= 0;
yma:= 0; ymi:= 0;  y:= 0;
xma:= 0; xmi:= 0;  x:= 0;
For i:= 1 to L do
 begin
  Case s[i] of
    '1': begin Dec(x); Dec(y); end;
    '2': Dec(y);
    '3': begin Inc(x); Dec(y); end;
    '4': Inc(x);
    '5': begin Inc(x); Inc(y); end;
    '6': Inc(y);
    '7': begin Dec(x); Inc(y); end;
    '8': Dec(x);
  end;
  If y > yma Then yma:= y;
  If y < ymi Then ymi:= y;
  If x > xma Then xma:= x;
  If x < xmi Then xmi:= x;
 end;
tb:= yma + Abs(ymi);
y:= (n - tb) div 2 + Abs(ymi);
lr:= xma + Abs(xmi);
x:= (m - lr) div 2;
MPosl[x, y]:= 1;
For i:= 1 to L do
 begin
  Case s[i] of
    '1': begin Dec(x); Dec(y); end;
    '2': Dec(y);
    '3': begin Inc(x); Dec(y); end;
    '4': Inc(x);
    '5': begin Inc(x); Inc(y); end;
    '6': Inc(y);
    '7': begin Dec(x); Inc(y); end;
    '8': Dec(x);
  end;
  If (x > n - 1) or (y > m - 1) or (x < 0) or (y < 0) Then
    begin
      spr:= 'The coordinates go beyond the boundaries of the
```

```
array! x = ` + IntToStr(x) + `, y = ` + IntToStr(x) + `!';
    If Application.MessageBox(PChar(spr),
                 'ALERT!', MB_OKCANCEL + MB_DEFBUTTON1 + MB_
ICONERROR) = IDCANCEL
        Then Exit;
    end
  else MPosl[x, y]:= 1;
 end;
end;
```

Listing of the DrawF embedded procedure is:

```
procedure DrawF(im: TImage; MM: Variant);
var i, j: Integer;
begin
If Fl_BMP Then
 begin
  For i:= 0 to n - 1 do
   For j:= 0 to m - 1 do
     begin
       If MM[i, j] = 1 Then im.Canvas.Pixels[i, j]:= clBlack else
im.Canvas.Pixels[i, j]:= clWhite;
     end;
 end
 else
 begin
  im.Canvas.Brush.Color:= clBlue;
  For i:= 0 to n - 1 do
   For j:= 0 to m - 1 do
     begin
       If MM[i, j] = 1 Then im.Canvas.Rectangle(i * 10 + 1, j *
10 + 1, i * 10 + 9, j * 10 + 9);
     end;
 end;
end;
```

The listing of the embedded procedure for fixing single errors has the form:

```
Procedure Ispr_Odin_Osh(var KodIsh: TKodAr);
var i, j, ind_Pred, ind, ind_Sled, ind_2Sled: Integer;
    L_1, LProm, n:Integer;
    Fl_Line:Boolean;
begin
L_1:= Length(KodIsh) - 1;
LProm:= L_1;
For i:= 0 to L_1 do
 begin
```

```
  Application.ProcessMessages;
  If i > LProm Then Break;
  If i = 0 Then begin ind_Pred:= L_1;  ind:= i;  ind_Sled:= i +
1;  ind_2Sled:= i + 2;  end;
  If i = LProm Then begin ind_Pred:= i - 1; ind:= i; ind_Sled:=
0; ind_2Sled:= 1;  end;
  If i = LProm - 1 Then begin ind_Pred:= i - 1; ind:= i; ind_
Sled:= L_1; ind_2Sled:= 0;  end;
  If (i > 0) and (i < LProm - 1) Then begin ind_Pred:= i - 1;
ind:= i; ind_Sled:= i + 1; ind_2Sled:= i + 2;  end;
  n:= Abs(KodIsh[ind].Y - KodIsh[ind_Sled].Y);
  Fl_Line:= n in [2, 4, 6];
  Case KodIsh[ind].X of
   1:If (KodIsh[ind_Sled].X = 1) and (KodIsh[ind_Pred].Y =
KodIsh[ind_2Sled].Y) and Fl_Line Then
       begin
        If KodIsh[ind_Pred].Y mod 2  = 0 Then n:= 2 else n:= 1;
        If (KodIsh[ind_Pred].Y mod 2  = 0) and
(Abs(KodIsh[ind].Y - KodIsh[ind_Sled].Y) = 4)
          Then n:= 0;
        KodIsh[ind_Pred].X:= KodIsh[ind_Pred].X +
KodIsh[ind_2Sled].X + n;
        If i = LProm Then
         begin
          For j:= 0 to LProm - 3 do
           begin
            KodIsh[j].X:= KodIsh[j + 2].X;
            KodIsh[j].Y:= KodIsh[j + 2].Y;
           end
         end
          else
           For j:= i to LProm do
            begin
             KodIsh[j].X:= KodIsh[j + 3].X;
             KodIsh[j].Y:= KodIsh[j + 3].Y;
            end;
        LProm:= LProm - 3;
        SetLength(KodIsh, LProm + 1);
       end;
   end;
  end;
end;
```

The listing of the embedded calculation procedure for all the turns and shifts options has the form:

```
Procedure VseVarianty(KodPosle: TKodAr; var VseVa:
TMasPointPrim);
```

```
var KodPr, KodPr1: TKodAr;
    i, j, L, j1, LMas: Integer;
begin
LMas:= 1;
SetLength(VseVa, LMas);
L:= Length(KodPosle);
SetLength(VseVa[0].MasPoint, L);
SetLength(KodPr, L);
SetLength(KodPr1, L);
For i:= 0 to L - 1 do
 begin
  VseVa[0].MasPoint[i].X:= KodPosle[i].X;
  VseVa[0].MasPoint[i].Y:= KodPosle[i].Y;
 end;
VseVa[0].Prim:= 'Full match';
For i:= 0 to L - 1 do
 begin
  KodPr[i].X:= KodPosle[i].X;
  KodPr[i].Y:= KodPosle[i].Y;
 end;
For i:= 1 to 7 do
 begin
  LMas:= LMas + 1;
  SetLength(VseVa, LMas);
  SetLength(VseVa[LMas - 1].MasPoint, L);
  Povorot45KodAr(KodPr, KodPr);
  For j:= 0 to L - 1 do
   begin
    VseVa[LMas - 1].MasPoint[j].X:= KodPr[j].X;
    VseVa[LMas - 1].MasPoint[j].Y:= KodPr[j].Y;
   end;
  VseVa[LMas - 1].Prim:= 'Rotated to' + IntToStr(i * 45) + '°';
  For j1:= 0 to L - 1 do
   begin
    KodPr1[j1].X:= KodPr[j1].X;
    KodPr1[j1].Y:= KodPr[j1].Y;
   end;
  For j:= 1 to L do
   begin
    LMas:= LMas + 1;
    SetLength(VseVa, LMas);
    SetLength(VseVa[LMas - 1].MasPoint, L);
    Sdvig_Left(KodPr1, KodPr1);
    For j1:= 0 to L - 1 do
     begin
      VseVa[LMas - 1].MasPoint[j1].X:= KodPr1[j1].X;
      VseVa[LMas - 1].MasPoint[j1].Y:= KodPr1[j1].Y;
     end;
    VseVa[LMas - 1].Prim:= 'Rotated to'' + IntToStr(i * 45) +
```

```
'°, shifted ' + IntToStr(j) + ' time';
   end;
 end;
 VseVar_String(VseVa);
end;
```

The listing of the embedded match check procedure has the form:

```
procedure Sravnenie(karIsh, karDict: TKodAr; var FlSov,
FlPolSov: Boolean);
var LIsh, LDict, i, ks, knes, kos, kones, kXs, kXns: Integer;
    ArrOtnIsh, ArrOtnDict: ArrOtn;
    delta: Real;
begin
FlSov:= False; FlPolSov:= False;
ks:= 0;   knes:= 0;
kos:= 0;   kones:= 0;   kXs:= 0;   kXns:= 0;
LIsh:= Length(karIsh);
LDict:= Length(karDict);
delta:= StrToFloat(glavn.LabeledEdit2.Text);
If LIsh = LDict Then
 begin
  For i:= 0 to LIsh - 1 do
   begin
    If karIsh[i].Y = karDict[i].Y Then Inc(ks) else Inc(knes);
    If karIsh[i].X = karDict[i].X Then Inc(kXs) else Inc(kXns);
   end;
  If (knes = 0)and(kXns = 0)Then begin FlPolSov:= True; Exit;
end;
  If (knes = 0)and(kXns = 0) Then
   begin
    Otn(karIsh, ArrOtnIsh);
    Otn(karDict, ArrOtnDict);
    For i:= 0 to Length(ArrOtnIsh) - 1 do
      If Abs(ArrOtnIsh[i] - ArrOtnDict[i]) < delta Then
Inc(kos) else Inc(kones);
    If kones = 0 Then FlSov:= True;
   end;
 end;
end;
```

Listing of the ParNap procedure:

```
Procedure Tglavn.ParNap(KAO:TKodAr;var MPNa:TMasParNap);
var
  i, j, l, c, k,StNaI: Integer;
  Ch:  Char;
```

```
 StNa: String[2];
 fpn: TextFile;
 naf, snapr: String;
 MPNaIsh: TMasParNap;
 NameNap, Nap: Byte;
 a, b: Integer;
Begin
 snapr:= Point_Posl(KAO);
 naf:= CurDir+'углы\углы.txt';
 AssignFile(fpn,naf);
 If FindFirst(naf,faArchive,SR) = 0 Then
   begin
    Reset(fpn);
    k:= 0;
    while not Eof(fpn) do
     begin
      Read(fpn,NameNap,a,b,Nap);
      SetLength(MPNaIsh,k+1);
      MPNaIsh[k].name:= NameNap;
      MPNaIsh[k].a:= a;
      MPNaIsh[k].b:= b;
      MPNaIsh[k].napr:= Nap;
      Inc(k);
     end;
   end;
 l:= Length(snapr);
 Ch:= snapr[1];
 c:= 1; j:= 0;
 For i:= 1 to l do
  begin
   If (snapr[i] = Ch) Then Inc(c)
    else
     begin
      StNa:= Ch+snapr[i];
      StNaI:= StrToInt(StNa);
      for k:=  0 to Length(MPNaIsh) - 1 do
       if MPNaIsh[k].name=StNaI then
        begin
         SetLength(MPNa, j+1);
         MPNa[j]:= MPNaIsh[k];
         Inc(j);
         Ch:= snapr[i];
         Break;
        end;
     end;
   If i=l Then
    If (snapr[i] = snapr[1]) Then Inc(c)
     else
      begin
```

```
        StNa:= snapr[i]+snapr[1];
        StNaI:= StrToInt(StNa);
        for k:=  0 to Length(MPNaIsh) - 1 do
         if MPNaIsh[k].name=StNaI then
          begin
            SetLength(MPNa, j+1);
            MPNa[j]:= MPNaIsh[k];
            Break;
          end;
     end;
   end;
end;
```

Chapter 8
New Evolutionary Model of Life Based on Cellular Automata

ABSTRACT

The chapter presents software that implements models of asynchronous cellular automata with a variable set of active cells. The software is considering one of the modifications of the game Conway "Life". In the proposed model "New Life," the possibility of functioning of a separate "living" cell is realized, which, when meeting with other "living" cells, participates in the "birth" of new "living" cells with a different active state. Each active state is determined by a code that is formed by the state values of the cells of the neighborhood. Variants of the evolution of the universe based on the surroundings of von Neumann and Moore are considered. This program uses restrictions on the number of "born" cells in order to limit the overpopulation of the universe. Possible goals and objectives to be solved in the use of "New Life" are also considered.

INTRODUCTION

The chapter presents software that implements models of asynchronous cellular automata with a variable set of active cells. The software is considering one of the modifications of the game Conway "Life". In the proposed model "New Life" the possibility of functioning of a separate "living" cell is realized, which, when meeting with other "living" cells, participates in the "birth" of new "living" cells with a different active state. Each active state is determined

DOI: 10.4018/978-1-7998-2649-1.ch008

by a code that is formed by the state values of the cells of the neighborhood. Variants of the evolution of the universe based on the surroundings of von Neumann and Moore are considered. This program uses restrictions on the number of "born" cells in order to limit the overpopulation of the universe. Possible goals and objectives to be solved in the use of "New Life" are also considered.

CONWAY GAME OF "LIFE"

The English mathematician John Conway invented game of Life in 1970. On the basis of a cellular automata, a universe is presented in which there are two types of cells ("living" and "dead"). Dead cells are represented by a logical "1" state. Eight cells are used to realize the neighborhood (Moore neighborhood).

At the initial moment of the game of life is carried out by the initial filling of all CA cells with "living" and "dead" cells. That is, initially the corresponding cells are set to logical states "1" and "0". At each time step, a new generation is calculated. To do this, the following rules are used:

- If a cell has a logical "0" state ("dead" cell) and among neighboring cells three cells have a logical "1" state, the cell goes into a logical "1" state (life is born).
- If a cell has a state of logical "1" (a "living" cell) and among its neighboring cells there are two or three cells having a state of a logical "1" (("live" cells), then the cell remains in a state of logical "1" (continues live).
- If a cell has a logical "1" state (a "living" cell) and among its neighbors there are less than two or more than three cells that have a logical "1" state, then the cell goes into a logical "0" state ("dies").

The game is terminated according to the following rules.

- All cells go into a logical "0" state ("die").
- At one of the time steps, the state of the CA coincides with one of the states of the CA at the previous time steps.
- At the next time step, none of the CA cells changes their state.

The first game of life was published in (Gardner 1970). In fact, the game of life comes down to setting the initial state (first population), which during a certain circle of time steps will give the necessary evolution of CA. As a result of the studies, stable forms were obtained that have the following classification.

- Stable figures are figures that do not change their shape at each subsequent time step.
- Long-livers are figures that take stable forms through a long number of time steps.
- Periodic figures are figures whose forms are periodically repeated after a certain number of time steps.
- Moving figures are figures whose state repeats after a certain displacement.
- Shotguns are repeating figures, but a moving figure appears.
- Steam locomotives are moving identical shapes that are left by other permanent or periodic shapes.
- Eaters are permanent shapes that remain in collision with other shapes.

Currently, there are studies of various modifications of the game of life (Adamatzky 2010; Adamatzky, 2018; Komosinnski, & Adamatzky 2009;10th International Conference on Cellular Automata for Research and Industry [ACRI 2012]; 11th International Conference on Cellular Automata for Research and Industry [ACRI 2014]; 12th International Conference on Cellular Automata for Research and Industry, [ACRI 2016]; 13th International Conference on Cellular Automata for Research and Industry, [ACRI 2018]).

There is also the inverse problem of Conway, which is to find the initial figure in the early time steps (Adamatzky 2010; Adamatzky, 2018; Komosinnski, & Adamatzky 2009).

An analysis of existing information sources showed that the game of life is well researched and research continues. A search is carried out for new forms that are formed by "living" cells in the CA field and which give the desired forms after a certain number of time steps. In many cases, software is implemented that allows each user to experiment with their own forms.

Although the game is called "Life", however, it does not describe real life, nor does it describe a model for the interaction of living individuals with each other. Therefore, this chapter is aimed at modeling the processes of interaction of living organisms with the closest proximity to the process of evolution in nature.

To achieve this goal, ACA with active cells are used, the structures of which are described in detail in previous chapters.

EVOLUTION MODEL OF THE INTERACTION OF LIVING ORGANISMS BASED ON ACA

In the previous section of the chapter, the interaction of CA cells was described on the basis of several simple rules, which are determined by the functions of comparing the number of states of the neighborhood cells and the eigenstate. Based on these simple functions, the control cell changes its state or remains in the same state. Due to the established structure of CA, with each subsequent time step, the number of "living" cells may change, but may remain in the same amount.

It is not difficult to notice that the following limiting property is inherent in such a CA. At the initial moment of time, a neighborhood should be set, in which there should be a certain number of "living" cells. If this amount is not maintained, then the birth of a new cell is not carried out and the evolution process is completed. In the "life" game, "living" cells do not "move" along the CA field, which does not allow the appearance of "living" cells outside the colony of "living" cells. Life is born only in large colonies of "living" cells. In this case, the cell colony begins where it was originally assigned. In real life, this is not so. Colonies can exist in one place and give birth to living cells that spread throughout the universe and form new colonies of "living" cells. In Conway's game, "living" cells can only propagate provided that they propagate that the three extreme cells of the colony belong to the neighborhood of one of the cells that have a logical "0" state. A colony of living cells will spread towards this cell.

This chapter describes a new model of the spread of life in the universe based on the theoretical propositions proposed and discussed in previous chapters.

General Description of the New Evolution of Life

In Conway's game of "Life," cells with one property were considered, which consisted of finding a cell in a state of logical "0" or logical "1". That is, a cell can have only two states, which are the main states. The new model uses ACA, in which the cell can also be in one of two states (logical "0" or

"1"). These states are formed as a result of the execution of a given LSF. In addition, the cell may be in an additional active state. Moreover, there can be many active states.

The essence of implementing the evolutionary model of life is as follows.

At the initial moment of time, the cells of the ACA (the universe) are set to logical states "0" and "1". Moreover, all these cells are inactive ("dead").

At the next moment of time, cells that become active ("living") are selected in the ACA field (universe). These cells can be active and have one of the basic states of the logical "0" or "1". Each cell, being in an active state, performs a given LSF. The arguments of LSF are the states of the neighborhood of the active cell. At the initial moment, the shape of the neighborhood for each active cell is also set.

In addition, each active cell performs LTF, according to which a cell is selected near the active cell, which will become active in the next time step. In fact, LTF indicates the direction of transmission of the active signal. Different LTFs can be used to implement transitions of active cells. If all active cells with different active states perform the same LTF, then the LTF arguments must be different for each active state. Moreover, the values of the LTF arguments for each active state are taken from different sources. Each active state must have its own law of active state transfer. The coincidence of LTF and the sources of their arguments for different active states equalizes these active states.

Thus, at the initial moment of time, the following states of the universe (aperiodic cellular automaton) will be present.

1. All ACA cells are in the basic states of the logical "0" and "1". These states in the ACA field are distributed differently depending on the task of modeling.
2. Among all the cells of the universe there are "living" cells (active). These cells can have different active states (different active potential). Active cells are located in user-defined ACA cells.
3. Each cell in the universe can perform a given LSF when it enters an active state.
4. Each cell is ready for transition to one of the active states.
5. Each active cell performs LTF, which indicates the direction of transmission of the active signal.

After completion of the initial settings, the interaction of "living" cells at each time step is calculated. At each time step, the following events occur.

- One of the cells in the neighborhood of each active cell goes into the active state of the previous neighboring active cell goes into, and the previous active cell "dies" (goes into an inactive state). The new active cell is selected from the neighborhood cells according to LTF, which the active cell performs in the previous time step.
- Active cells perform LSF, according to which they can change their basic information state.
- If several active ("living") cells with different active states are in close interaction with each other, then they form a new active cell (a new "living" cell is born) with a new active state. A new active state is formed in accordance with the given rules of evolution.
- If two active cells with identical active states are in close interaction, then these cells "die" and the "new life" does not "born" in the universe, and the active interacting cells do not "die".
- If two active cells are in close interaction and do not fall under any rules of evolution, then the "new life" is not "born", and the active interacting cells do not "die".

The developer may change the described conditions and rules during the construction of the model. For example, a separate LSF and LTF may be filed for each active state. Close interaction rules can also be changed. Finally, the rules are described for invariable states of inactive cells in the universe.

The following sections will describe several options for the game "new life", as well as software implementation.

SOFTWARE IMPLEMENTATION OF THE NEW LIFE MODEL BASED ON ACA WITH VON NEUMANN NEIGHBORHOOD

The use of the von Neumann neighborhood significantly limits the number of active signal transmission directions. An active signal can be transmitted in one of four directions (right, left, up, down). Four can be specified by two bit binary code. This code can be generated by two cells in the neighborhood. The fifth chapter examined the coding of each active state. Table 1, which presents the codes of the "birth" of new active cells, is also presented there. Moreover, all possible active states are set by the table, which eliminates the chance of the appearance of a new active cell. If inactive cells do not perform LSF, then the use of coding according to table 1 leads to cycles with a small number of active cells. An example of the functioning of such an ACA on Figure 1 is presented.

Figure 1. An example of ACA functioning using the von Neumann neighborhood

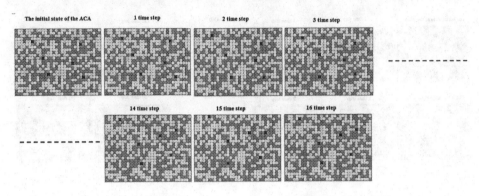

In this example, active cells perform the XOR function as LSF, and active states are encoded according to Table 1. At 14, 15 and 16 time steps cycles are shown that do not change evolution at further time steps. Each active cell is highlighted in red, and two pink cells next to each red indicate the code of the active state and state of the LTF arguments of the active cell.

An example clearly shows that in such states of ACA, active cells quickly enter the cycle and no further changes occur. In order to prevent this from happening, it is necessary to introduce the following additional functions.

- Choosing a LTF that does not give cycles;
- The choice of LSF, which changes the state of the cell so that the cell does not return to this state in the next few time steps;
- Each cell of the universe (living and dead) performs LSF at every time step;
- Change in the state of cells in the neighborhood of the active cell at given steps;
- Change in the active state of each cell after a certain number of time steps.

Each of these points is a task that must be solved and possibly solved using special methods and algorithms. For the software implementation under consideration, the LTF described in the fifth chapter is used. The example presented in Figure 1 uses the XOR function as an LSF. Figure 2 shows the evolution of the universe for the AND and OR functions that are used as LSFs.

Figure 2. Examples of universe evolution for AND and OR functions that use as LSF

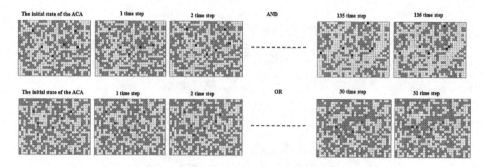

For an OR function, the universe is filled with cells in a logical "1" state. Six active cells with different active states were given. Several active cells move without going into a cycle. At the 28th time step, two active cells interact in one cell and at the 30th time step the cells are separated, and at the 31st time step they "die". Four active cells remain.

Six active cells with different active states are also specified for AND function. Several cells enter the cycle already at the first time steps. However, there are active cells that continue to move. At 142 time steps, two cells interact and all interacting new active cells move in the same direction according to the encoding. At the 145 time step, only four active cells remain that fall into the cycle and changes in evolution end.

If active cells perform a XOR function, then with a small number of active cells, cycles occur after a small number of time steps.

There is also the ability to specify different LSFs for active cells. An example of the evolution of the universe when using different LSFs on Figure 3 is presented.

Figure 3. An example of the evolution of the universe for the initial four active cells that perform XOR, AND and OR functions as LSF

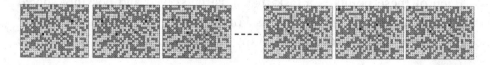

As can be seen from the presented example, with a small number of active cells, all active cells enter the cycle through a small number of time steps.

These examples have been considered for a consistent environment. Not all "dead" cells changed their state at each time step of evolution. If we assume that the universe changes with each time step, then all inactive ACA cells at each time step can change their basic information state in accordance with the selected LSF. In this situation, active cells do not always enter the cycle. Several modes of operation of such an ACA are possible.

- Active and inactive cells perform the same LSF.
- Active and inactive cells perform different LSFs.

In the first mode, the states of all cells change identically. However, active cells also perform LTF and can interact with each other. In this case, new active cells may appear or old ones may disappear. An example of the evolution of the universe in this mode on Figure 4 is presented.

Figure 4. An example of the evolution of the universe in the mode of its change at each time step

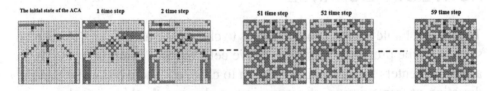

In Figure 4, at each time step, all cells in the universe perform an XOR operation. As you can see, the universe has a cycle of 8 time steps in length (starts at 51 time steps and repeats at 59 time steps). ACA states at 51 and 59 time steps coincide.

The second option is characterized by inhomogeneity of active cells, which is in another LSF, which differs from inactive cells. In this case, active cells "move" along the ACA field. An example of the evolution of the universe in the second mode on Figure 5 is presented.

Active cells perform another LSF and transmit active signals to neighboring cells. Also, active cells can interact with each other and form new active cells. In this example, all the "dead" cells of the universe perform XOR functions,

Figure 5. An example of the evolution of the universe where LSFs of "living" cells differ from LSFs of "dead" cells

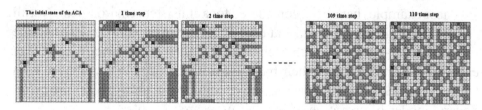

and active cells perform the AND function (3 active cells are highlighted in green) and OR function (3 active cells are highlighted in blue).

An example shows that there are no cycles and the universe changes at each time step. At a certain time step of evolution, colonies of "living" cells are formed, which can disappear quickly, and a large number of time steps can last. This example used the same initial settings as in the example shown in Figure 4. The second option is more effective. In such a combination LSF (AND and OR). Several time steps for such movement in Figure 6 are presented.

ACTIVE STATE CHANGE

The most reliable way to avoid cycles is to change the active state. As can be seen from the previous examples, active cells quickly enter the cycle. If an active cell enters the cycle, it is enough to change its active state so that the direction of transmission of active signals changes. In this case, the active cell must determine that it is in the cycle, which complicates the cell itself. An additional analysis of the state of the neighborhood cells at the previous time steps is required. The best option is to choose the number of time steps through which the cell will change its active state, regardless of the onset of the cycle. Such a number of time steps is determined as a result of studying

Figure 6. Example of pairwise movement of active cells with different LSFs

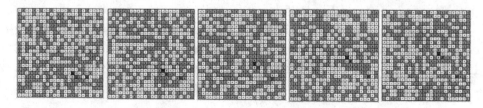

the evolution of ACA behavior in various states. An example of the evolution of ACA, when active cells change their state after 3, 5 10 time steps, on Figure 7 is presented.

Figure 7. The evolution of the universe when active cells change their active state

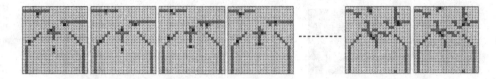

From this example it is seen that the "movement" of active cells is different for a different number of steps of changing the active state. In this case, cycles are practically not observed. Colonies of new cells are also formed that function for a while, and then disappear. The general states of the universe are also different for each cycle of changes in the active state. No cycles observed.

In addition to cycles, there is another problem, which consists in the gradual overpopulation of the universe. Therefore, the program has the following restrictions.

- new active cells will not be created in an active cell that already contains two or more active functions (limiting the unjustified growth in the number of active functions that may be present in one cell);
- when the CA is operating in the neighborhood of Moore, the interaction of active cells is possible only at the intersection of 1 of the 3 cells of the transition mask (an attempt to balance the number of deleted and created CA - after all, deletion occurs at the intersection of 3 out of 3, and the creation of a new one is in all other cases of intersection, i.e. 1 out of 3 and 2 out of 3, which at least doubles the probability of unlimited reproduction of active cells). When the full interaction mode is activated, an avalanche-like explosive increase in the number of active cells is almost always observed when a certain critical number of cells is reached;
- prevention of the possible onset of an "epidemic" of active cells when their number exceeds half the number of all CA cells.

An example of the evolution of ACA with the full interaction of active cells (without limitation) on Figure 8 is presented.

Figure 8. An example of the evolution of the universe without restrictions on the "birth" of new active cells

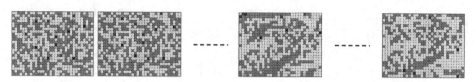

If there are no restrictions, then a quick overpopulation of the universe with active cells is possible. However, such overpopulation rarely occurs for the von Neumann neighborhood, since the destruction of active cells with identical active states often occurs.

An example of the evolution of the universe with established fertility restrictions under the same initial conditions on Figure 9 is presented.

Figure 9. An example of the evolution of the universe with birthrate restrictions for new active cells

In this example, the number of active cells is less at the same time steps as in the previous example. This situation avoids overpopulation of the universe. New active cells do not appear when several active cells interact. The von Neumann neighborhood allows a clear and understandable study of the behavior of such ACAs, since the number of analyzed cells is 4, and a small number of states are formed using 4 cells. However, the von Neumann neighborhood significantly limits the behavior of ACA. Therefore, developers are striving to increase the number of cells that form the neighborhood. In this regard, the most popular is the neighborhood of Moore.

Table 1. Table of codes of active cells in the interaction of existing active cells

Codes of Active States of Interacting Cells				New Active State	
First Active Cell		Second Active Cell		Active State Number	Active Status Code
Active State Number	Active Status Code	Active State Number	Active Status Code		
1	$a_1\,a_2\,a_3$	1	$a_1\,a_2\,a_3$		0
	$a_1\,a_2\,a_3$	2	$a_1\,a_2\,a_4$		$a_3\,a_4$
	$a_1\,a_2\,a_3$	3	$a_1\,a_2\,a_5$		$a_3\,a_5$
	$a_1\,a_2\,a_3$	4	$a_1\,a_2\,a_6$		$a_3\,a_6$
	$a_1\,a_2\,a_3$	5	$a_1\,a_2\,a_7$		$a_3\,a_7$
	$a_1\,a_2\,a_3$	6	$a_1\,a_2\,a_8$		$a_3\,a_8$
	$a_1\,a_2\,a_3$	7	$a_1\,a_3\,a_4$		$a_2\,a_4$
	$a_1\,a_2\,a_3$	8	$a_1\,a_3\,a_5$		$a_2\,a_5$
	$a_1\,a_2\,a_3$	9	$a_1\,a_3\,a_6$		$a_2\,a_6$
	$a_1\,a_2\,a_3$	10	$a_1\,a_3\,a_7$		$a_2\,a_7$
	$a_1\,a_2\,a_3$	11	$a_1\,a_3\,a_8$		$a_2\,a_8$
	$a_1\,a_2\,a_3$	12	$a_1\,a_4\,a_5$		$a_2\,a_3\,a_4\,a_5$
	$a_1\,a_2\,a_3$	13	$a_1\,a_4\,a_6$		$a_2\,a_3\,a_4\,a_6$
	$a_1\,a_2\,a_3$	14	$a_1\,a_4\,a_7$		$a_2\,a_3\,a_4\,a_7$
	$a_1\,a_2\,a_3$	15	$a_1\,a_4\,a_8$		$a_2\,a_3\,a_4\,a_8$
	$a_1\,a_2\,a_3$	16	$a_1\,a_5\,a_6$		$a_2\,a_3\,a_5\,a_6$
	$a_1\,a_2\,a_3$	17	$a_1\,a_5\,a_7$		$a_2\,a_3\,a_5\,a_7$
	$a_1\,a_2\,a_3$	18	$a_1\,a_5\,a_8$		$a_2\,a_3\,a_5\,a_8$
	$a_1\,a_2\,a_3$	19	$a_1\,a_6\,a_7$		$a_2\,a_3\,a_6\,a_7$
	$a_1\,a_2\,a_3$	20	$a_1\,a_6\,a_8$		$a_2\,a_3\,a_6\,a_8$
	$a_1\,a_2\,a_3$	21	$a_1\,a_7\,a_8$		$a_2\,a_3\,a_7\,a_8$
	$a_1\,a_2\,a_3$	22	$a_2\,a_3\,a_4$		$a_1\,a_4$
	$a_1\,a_2\,a_3$	23	$a_2\,a_3\,a_5$		$a_1\,a_5$
	$a_1\,a_2\,a_3$	24	$a_2\,a_3\,a_6$		$a_1\,a_6$
	$a_1\,a_2\,a_3$	25	$a_2\,a_3\,a_7$		$a_1\,a_7$
	$a_1\,a_2\,a_3$	26	$a_2\,a_3\,a_8$		$a_1\,a_8$
	$a_1\,a_2\,a_3$	27	$a_2\,a_4\,a_5$		$a_1\,a_3\,a_4\,a_5$
	$a_1\,a_2\,a_3$	28	$a_2\,a_4\,a_6$		$a_1\,a_3\,a_4\,a_6$
	$a_1\,a_2\,a_3$	29	$a_2\,a_4\,a_7$		$a_1\,a_3\,a_4\,a_7$
1	$a_1\,a_2\,a_3$	30	$a_2\,a_4\,a_8$		$a_1\,a_3\,a_4\,a_8$
	$a_1\,a_2\,a_3$	31	$a_2\,a_5\,a_6$		$a_1\,a_3\,a_5\,a_6$
	$a_1\,a_2\,a_3$	32	$a_2\,a_5\,a_7$		$a_1\,a_3\,a_5\,a_7$
	$a_1\,a_2\,a_3$	33	$a_2\,a_5\,a_8$		$a_1\,a_3\,a_5\,a_8$
	$a_1\,a_2\,a_3$	34	$a_2\,a_6\,a_7$		$a_1\,a_3\,a_6\,a_7$
	$a_1\,a_2\,a_3$	35	$a_2\,a_6\,a_8$		$a_1\,a_3\,a_6\,a_8$
	$a_1\,a_2\,a_3$	36	$a_2\,a_7\,a_8$		$a_1\,a_3\,a_7\,a_8$
	$a_1\,a_2\,a_3$	37	$a_3\,a_4\,a_5$		$a_1\,a_2\,a_4\,a_5$
	$a_1\,a_2\,a_3$	38	$a_3\,a_4\,a_6$		$a_1\,a_2\,a_4\,a_6$
	$a_1\,a_2\,a_3$	39	$a_3\,a_4\,a_7$		$a_1\,a_2\,a_4\,a_7$
	$a_1\,a_2\,a_3$	40	$a_3\,a_4\,a_8$		$a_1\,a_2\,a_4\,a_8$
	$a_1\,a_2\,a_3$	41	$a_3\,a_5\,a_6$		$a_1\,a_2\,a_5\,a_6$
	$a_1\,a_2\,a_3$	42	$a_3\,a_5\,a_7$		$a_1\,a_2\,a_5\,a_7$
	$a_1\,a_2\,a_3$	43	$a_3\,a_5\,a_8$		$a_1\,a_2\,a_5\,a_8$
	$a_1\,a_2\,a_3$	44	$a_3\,a_6\,a_7$		$a_1\,a_2\,a_6\,a_7$
	$a_1\,a_2\,a_3$	45	$a_3\,a_6\,a_8$		$a_1\,a_2\,a_6\,a_8$
	$a_1\,a_2\,a_3$	46	$a_3\,a_7\,a_8$		$a_1\,a_2\,a_7\,a_8$
	$a_1\,a_2\,a_3$	47	$a_4\,a_5\,a_6$		$a_1\,a_2\,a_3\,a_4\,a_5\,a_6$
	$a_1\,a_2\,a_3$	48	$a_4\,a_5\,a_7$		$a_1\,a_2\,a_3\,a_4\,a_5\,a_7$
	$a_1\,a_2\,a_3$	49	$a_4\,a_5\,a_8$		$a_1\,a_2\,a_3\,a_4\,a_5\,a_8$
	$a_1\,a_2\,a_3$	50	$a_4\,a_6\,a_7$		$a_1\,a_2\,a_3\,a_4\,a_6\,a_7$
	$a_1\,a_2\,a_3$	51	$a_4\,a_6\,a_8$		$a_1\,a_2\,a_3\,a_4\,a_6\,a_8$
	$a_1\,a_2\,a_3$	52	$a_4\,a_7\,a_8$		$a_1\,a_2\,a_3\,a_4\,a_7\,a_8$
	$a_1\,a_2\,a_3$	53	$a_5\,a_6\,a_7$		$a_1\,a_2\,a_3\,a_5\,a_6\,a_7$
	$a_1\,a_2\,a_3$	54	$a_5\,a_6\,a_8$		$a_1\,a_2\,a_3\,a_5\,a_6\,a_8$
	$a_1\,a_2\,a_3$	55	$a_5\,a_7\,a_8$		$a_1\,a_2\,a_3\,a_5\,a_7\,a_8$
	$a_1\,a_2\,a_3$	56	$a_6\,a_7\,a_8$		$a_1\,a_2\,a_3\,a_6\,a_7\,a_8$

Table 2. A coding system for new active states according to the Moore neighborhood

Codes of Active States of Interacting Cells				New Active State	
First Active Cell		Second Active Cell		Active State Number	Active Status Code
Active State Number	Active Status Code	Active State Number	Active Status Code		
1	$a_1 a_2 a_3$	1	$a_1 a_2 a_3$		0
		2	$a_1 a_2 a_4$	37	$a_3 a_4 a_5$
		3	$a_1 a_2 a_5$	41	$a_3 a_4 a_6$
		4	$a_1 a_2 a_6$	38	$a_3 a_4 a_6$
		5	$a_1 a_2 a_7$	42	$a_3 a_4 a_7$
		6	$a_1 a_2 a_8$	40	$a_3 a_4 a_8$
		7	$a_1 a_3 a_4$	27	$a_2 a_4 a_5$
		8	$a_1 a_3 a_5$	27	$a_2 a_4 a_5$
		9	$a_1 a_3 a_6$	28	$a_2 a_4 a_6$
		10	$a_1 a_3 a_7$	29	$a_2 a_4 a_7$
		11	$a_1 a_3 a_8$	30	$a_2 a_4 a_8$
		12	$a_1 a_4 a_5$	56	$a_6 a_7 a_8$
		13	$a_1 a_4 a_6$	55	$a_5 a_7 a_8$
		14	$a_1 a_4 a_7$	54	$a_5 a_6 a_8$
		15	$a_1 a_4 a_8$	53	$a_5 a_6 a_7$
		16	$a_1 a_5 a_6$	52	$a_4 a_7 a_8$
		17	$a_1 a_5 a_7$	51	$a_4 a_6 a_8$
		18	$a_1 a_5 a_8$	50	$a_4 a_6 a_7$
		19	$a_1 a_6 a_7$	49	$a_4 a_5 a_8$
		20	$a_1 a_6 a_8$	48	$a_4 a_5 a_7$
		21	$a_1 a_7 a_8$	47	$a_4 a_5 a_6$
		22	$a_2 a_3 a_4$	12	$a_1 a_4 a_5$
		23	$a_2 a_3 a_5$	12	$a_1 a_4 a_5$
		24	$a_2 a_3 a_6$	13	$a_1 a_4 a_6$
		25	$a_2 a_3 a_7$	14	$a_1 a_4 a_7$
		26	$a_2 a_3 a_8$	15	$a_1 a_4 a_8$
		27	$a_2 a_4 a_5$	56	$a_6 a_7 a_8$
		28	$a_2 a_4 a_6$	55	$a_5 a_7 a_8$
		29	$a_2 a_4 a_7$	54	$a_5 a_6 a_8$
		30	$a_2 a_4 a_8$	53	$a_5 a_6 a_7$
		31	$a_2 a_5 a_6$	52	$a_4 a_7 a_8$
		32	$a_2 a_5 a_7$	51	$a_4 a_6 a_8$
		33	$a_2 a_5 a_8$	50	$a_4 a_6 a_7$
		34	$a_2 a_6 a_7$	49	$a_4 a_5 a_8$
		35	$a_2 a_6 a_8$	48	$a_4 a_5 a_7$
		36	$a_2 a_7 a_8$	47	$a_4 a_5 a_6$
		37	$a_3 a_4 a_5$	56	$a_6 a_7 a_8$
		38	$a_3 a_4 a_6$	55	$a_5 a_7 a_8$
		39	$a_3 a_4 a_7$	54	$a_5 a_6 a_8$
		40	$a_3 a_4 a_8$	53	$a_5 a_6 a_7$
		41	$a_3 a_4 a_6$	52	$a_4 a_7 a_8$
		42	$a_3 a_5 a_7$	51	$a_4 a_6 a_8$
		43	$a_3 a_5 a_8$	50	$a_4 a_6 a_7$
		44	$a_3 a_6 a_7$	49	$a_4 a_5 a_8$
		45	$a_3 a_6 a_8$	48	$a_4 a_5 a_7$
		46	$a_3 a_7 a_8$	47	$a_4 a_5 a_6$
		47	$a_4 a_5 a_6$	-------	-------
		48	$a_4 a_5 a_7$	-------	-------
		49	$a_4 a_5 a_8$	-------	-------
		50	$a_4 a_6 a_7$	-------	-------
		51	$a_4 a_6 a_8$	-------	-------
		52	$a_4 a_7 a_8$	-------	-------
		53	$a_5 a_6 a_7$	-------	-------
		54	$a_5 a_6 a_8$	-------	-------
		55	$a_5 a_7 a_8$	-------	-------
		56	$a_6 a_7 a_8$	-------	-------

NEW LIFE MODEL BASED ON ACA
WITH MOORE NEIGHBORHOOD

The Moore neighborhood allows you to increase the number of active states and the number of directions of transmission of the active signal. Since the Moore neighborhood consists of eight cells, three binary bits are used to encode them, which are formed at the outputs of three neighborhood cells. At the same time, several coding options can be used to form a new active cell. One of the encoding options according to the principle considered for the von Neumann neighborhood in Table 1 is presented.

As can be seen from Table 1, codes of new active states are formed, which consist of two and four binary bits. Since you cannot specify eight directions with a two-digit binary code (only four), these active states are not taken into account, and a new active cell is not "born". Also, active states represented by four-digit codes are not taken into account, since they can indicate a nonexistent cell in the neighborhood during the transmission of the active signal. Only states that are represented by three-digit binary code are used, and such active states cannot be obtained with this principle.

The program implements a coding system that generates states according to Table 2.

Table 4 describes the new states for one active state $a_1 a_2 a_3$. The coding of the new state is carried out in such a way that the code of the new active state is formed from three cells in the neighborhood of the coincident cell, which are not included in the code of any of the interacting active cells. Moreover,

Figure 10. An example of the formation of new active cells using the neighborhood of Moore

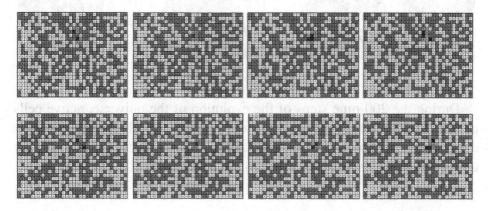

the union of neighborhood cells encoding the active states of two interacting active cells should include five neighborhood cells. An example of coding and "birth" of new active neighborhood cells on Figure 10 is presented.

The initial active cells are highlighted in red, next to which the neighborhood cells that encode the corresponding active state are highlighted in pink. After their interaction, a new active cell appears (highlighted in green), next to which three cells of its neighborhood are highlighted (highlighted in light green), which encode its active state according to Table 2. An example of the evolution of the universe without restrictions on overpopulation and based on the Moore neighborhood in Figure 11 is presented.

Figure 11. An example of the evolution of the universe without restrictions on overpopulation (Moore neighborhood)

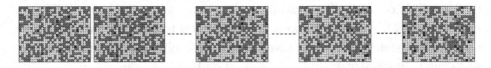

The number of active cells increases and the overpopulation of the universe is possible. If you establish the restrictions described earlier, then the balance of the population of the universe is respected. An example of the program operation taking into account restrictions on Figure 12 is presented.

Figure 12. An example of the evolution of the universe taking into account restrictions on overpopulation (using the Moore neighborhood)

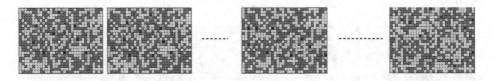

During the 200 time steps of the evolution of the universe, active cells interacted with each other. However, new cells did not form due to the limitations used. Using the Moore neighborhood allows prolonging the evolution of the universe without the onset of cycles of all active cells.

MODEL OF THE EVOLUTION OF THE UNIVERSE IN THE INTERACTION OF NEIGHBORING ACTIVE CELLS

The interaction of neighboring active cells that belong to the neighborhood of each active cell is described in chapter 5. In this case, two active cells interact with each other if each of these cells belongs to the neighborhood of the cell with which it interacts. All active cells use the same neighborhood. The software implementation uses the neighborhood of von Neumann and Moore.

Figure 13. Examples of interaction of neighboring cells in the neighborhood of von Neumann

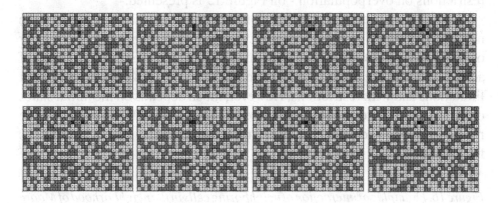

Figure 14. An example of the evolution of the universe in the interaction of neighboring active cells without restrictions on overpopulation (von Neumann neighborhood)

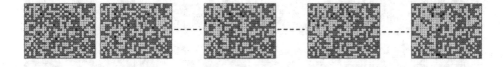

Figure 15. An example of the evolution of the universe in the interaction of neighboring active cells with restrictions on overpopulation (von Neumann neighborhood)

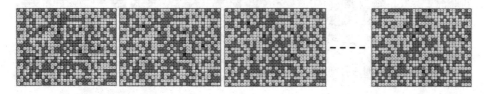

Model of the Evolution of the Universe Based on the Von Neumann Neighborhood

The program implements the interaction of neighboring active cells for the von Neumann neighborhood. An example of such an interaction on Figure 13 is presented.

The new active cell (highlighted in green) has a state that is encoded according to Table 1, and is also formed at the next time step in place of that interacting active cell, the active state code is larger than the active state code of another active cell. For such interaction, the program also has restrictions on overpopulation. An example of the complete interaction of active cells without restrictions on overpopulation in Figure 14 is presented, and with restrictions on overpopulation - on Figure 15 is presented.

At the 100 time step, the evolution of both universes is distinguished by the number of new active cells. However, at subsequent time steps, the number of active cells does not exceed half of the ACA cells. Formed colonies of active cells that are grouped in certain places in the universe, both without limitations and with restrictions. However, much depends on the initial states of the universe. For example, the number of new active cells in a universe with restrictions may exceed the number of new active cells without restrictions for different initial states.

Figure 16. Examples of interaction of neighboring cells in the neighborhood of Moore

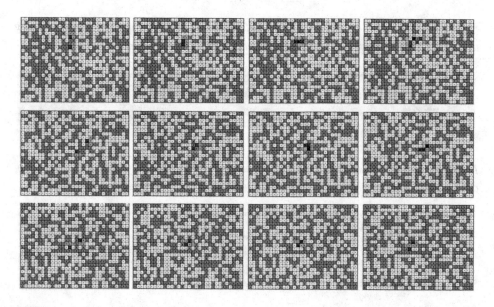

EVOLUTION MODEL BASED ON THE MOORE NEIGHBORHOOD

In the evolution of the universe based on the Moore neighborhood, the same restrictions are used as for the von Neumann neighborhood. Interaction of active cells using the Moore neighborhood for examples on Figure 16 is shown.

Other interaction options are considered here than shown in Figure 10. The variants of mismatch of the neighborhood cells encoding the active state are shown (a new active cell is not formed), as well as the coincidence variants of two neighborhood cells encoding the active states. An example of the evolution of the universe when there are no restrictions on overpopulation for the Moore neighborhood on Figure 17 is presented.

Figure 17. An example of the evolution of the universe without restriction on overpopulation (for the neighborhood of Moore)

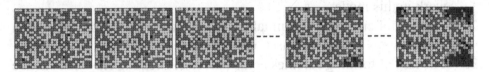

This example shows a total of 30 time steps in the evolution of the universe.

An example of the evolution of the universe with established restrictions on overpopulation on Figure 18 is presented.

Figure 18. An example of the evolution of a universe with overpopulation restrictions (for the Moore neighborhood)

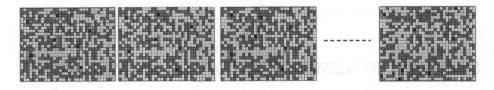

For this example, 100 time steps of evolution are considered. In a hundred time steps, one new active cell has appeared. The limited number of active states does not ensure the constant development of the universe and the emergence of new states. All states are limited to a coding table. Therefore, such a model requires further research in terms of life expectancy.

POSSIBLE TASKS AND OPTIONS FOR USING THE NEW LIFE MODEL

This section discusses the main options for the functioning of the "new life" model. The processes of interaction of active cells and the formation of new active cells are considered. If you use the software as a game, you can use the following game goals.

1. Interaction with the user in an interactive mode in order to prevent the meeting and interaction of active cells. During the game, it is necessary to change the state of the cells of the universe so that the active cells do not "move" to meet each other. The described program allows you to work in this mode.
2. The task of creating such initial states that gave the maximum duration of time steps before the onset of the general ACA cycle.
3. The task of forming colonies in the universe. Creating clusters of active cells.
4. The task of interaction between the colonies (unification, destruction, absorption).
5. The task of choosing the shape of the neighborhood and coding of active states to create a qualitatively new active state.

Having studied the proposed program, one can formulate new problems of modeling dynamic processes of interaction of objects. The proposed model can be used to model various physical processes, as well as the processes of interaction of biological organisms.

SOFTWARE IMPLEMENTATION OF THE NEW LIFE MODEL

Program Interface

The CA simulation program with a variable number of active cells (AC) is written in MS Visual C # 2008 using the working environment of the NET

Framework 2.0. The appearance of the program on Figure 19 is shown. The main form of the program consists of a tabbed area and a cellular automaton field. Between them is a counter of the current number of AC.

Figure 19. Appearance of the main form of the program

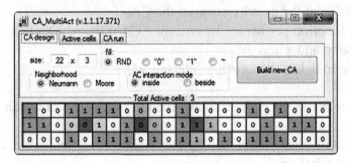

In Figure 19 the «CA design» tab, allowing to create a CA design is shown:

- its size;
- initial filling: randomly, all zeros, all ones, inversion of the current state;
- type of neighborhood for determining neighboring cells: according to Neumann and according to Moore;
- interaction mode of active cells: inside - internal coincidence in the coordinates of one cell, beside - coincidence in neighboring cells;
- button to build a new CA.

The "Active cell" and "CA run" tabs, which allow to set and control active cells and run simulation of the CA operation, respectively, on Figure 20 are shown.

Figure 20. Tabs for AC control and CA simulation

On the left is shown the tab "Active cell", on which:

- pointer to the coordinates of the currently selected active cell;
- a list showing a list of either all the active functions of the current cell, or a complete list of ACs (if AC is not selected);
- mask of the current AC code, which determines the value of its transition at the next iteration of the CA;
- local function performed by current AC (XOR, AND, OR);
- button to add a new AC;
- button to delete the existing selected AC.

In Figure 20 on the right is shown the tab "CA run", on which:

- simulation start mode: step-by-step manual and automatic, indicating the delay before the next iteration;
- a checkbox that allows to perform a cyclic shift clockwise (left) of the mask that defines the transition of AC. Helps prevent or significantly reduce cycling in CA operation with a small set of AC;
- checkbox that sets the mode of complete unlimited interaction of all ACs. If not checked, then the AC will meet the following restrictions: 1) new ACs will not be created in ACs that already contain two or more active functions (limiting the unjustified growth in the number of active functions that may be present in one cell), 2) when the CA is operating in the neighborhood of Moore, AC interaction is possible only at the intersection of 1 out of 3 cells of the transition mask (an attempt to balance the number of deleted and created CA - after all, deletion occurs at the intersection of 3 out of 3, and the creation of a new one - in all other cases of intersection, i.e. 1 out of 3 and 2 of 3, which at least doubles the probability of unlimited AC reproduction). When the full interaction mode is activated, an avalanche-like explosive increase in the number of ACs is almost always observed when a certain critical amount is reached;
- checkbox, which allows to prevent the possible onset of an "epidemic" of AC, when their number exceeds half the number of all CA cells;
- CA start / stop simulation button;
- concealment of the CA field. Useful for a large number of AKs to reduce interface rendering lags.

Some Features of Working With the Program

For the initial AC task, it is necessary to select the necessary cell on the CA field with the left mouse button, while holding Ctrl on the keyboard, and then specify the transition mask and local function. With the same selection (with Ctrl) of the existing AC, a list of its active functions, their masks and local functions will be shown. By double-clicking on the heading of the Active cell tab (or when switching to it from other tabs), a full list of ACs existing at this iteration step will be displayed.

Almost all controls are interactive, that is, they allow you to change the behavior and modes of operation of the CA on the go. That is, between iterations, you can change: the type of AC interaction (internal or adjacent AC), the values of specific cells and the entire state of the CA as a whole, add / remove ACs, view the list of active functions of individual ACs and display their complete list, control the mask cyclic shift, enable / disable full AC interaction, "epidemic" warning. Exceptions are the operations of creating a CA of a changed dimension and the choice of a different neighborhood - while all existing ACs are reset to zero and their initial list must be re-entered.

The Principle of Generating Parameters for a New AC

When working with CA in the neighborhood of Neumann, the new AC is set in full accordance with the calculations in Chapter 8, namely: the mask of the new AC is determined by two excluded mask cells when they intersect one cell (as you know, when they completely intersect 2 of 2 - mutual destruction AC, and with complete incompatibility - everything remains unchanged).

When working with CA in the neighborhood of Mura, a new AC, following the theory of Chapter 8, can appear when the masks of interacting ACs partially intersect. But since in this neighborhood the mask is no longer 2, but 3 cells, the appearance of a new AC will occur 2 times more often: at the intersection of 1 and at the intersection of 2 cells. In this case, the new mask for the first case (intersection one at a time) is defined as the three remaining uninhabited neighborhood cells, and in the second (intersection two), it is defined as two excluded mask cells and one lower-order cell of the unused neighborhood. But this interaction mode almost always ends with an explosive avalanche-like increase in the number of ACs, so the ability to limit AC growth only by crossing 1 mask cell from 3 is added to the program (CA run tab, checkbox "full interaction of AC"). The same checkbox (disabling it) limits the number of created active functions in one AC: if the cell in which a new AC is to be formed, there are already two or more active functions - creating a new AC is not performed. Thus, an AC can receive more than two functions only as a result of the transition of other ACs to it, but not by creating a new one, which also avoids an avalanche-like increase in the number of created ACs when a certain critical mass is reached.

A new local function is defined identically for all neighborhoods and types of AC interaction as the remainder of dividing by 3 sums of codes of local interacting functions AC + 1. Local function codes: XOR – 0, AND – 1, OR – 2. Then in the interaction LSF(1) = AND = 1 and LSF(2) = XOR = 0 we have: and LSF(3) = (1 + 0 + 1)%3 = 2 = OR

The Program Simulating the Work of the CA

The cellular automaton itself is represented by a two-dimensional array, the size of which can be arbitrarily adjusted using the "CA design" tab. Visualization of the cellular automaton is implemented by standard controls "label", also formed into a two-dimensional array of the appropriate size. When changing the size of the CA, all these arrays are dynamically recreated. The main code fragments describing the CA and its functions are presented in the Appendix.

The main intercellular interaction operations are performed using bit operations. So, for example, the state of all neighboring cells shown in Figure 21 matches (bitwise = number): $10100011_{(2)} = A3_{(16)} = 163_{(10)}$.

Figure 21. An example of the representation of neighboring cell states

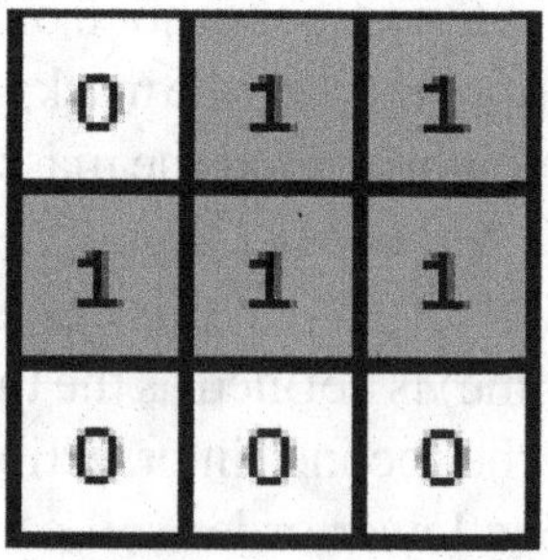

A complete list of ACs at each step of the CA iteration is stored in an array of such structures. Array size dynamically changes if necessary.

The modeling program is based on a miscalculation of one iteration of the CA, performed according to the following algorithm:

1. Search for active cells to be deleted that have the same mask of the displacement code, taking into account the currently selected interaction mode (inside, beside).
2. The total number of ACs becomes equal to their initial number minus the deleted ones in step 1.
3. For each AC, search for combinations of coincident / neighboring (depending on the interaction mode) ACs in which new ACs are possible.
4. For each new AC that appeared during step 3, its local function and mask of the displacement code are determined. The initial data for this are the corresponding parameters of the interacting active parent cells.
5. The total number of ACs increases by the number of added new ACs (steps 3-4).
6. For each AC, the value of its local function is calculated based on the values of neighboring cells.
7. For each AC, the direction of movement is calculated based on its mask of the code of movement and the values of neighboring cells.
8. All CA cells are assigned the corresponding values obtained in steps 6-7.
9. The end of the iteration, transition to step 1.

All points of the algorithm take into account neighborhood masks in the calculations. To perform bitwise operations, the Moore neighborhood detection mask will look like: $11111111_{(2)} = FF_{(16)}$; accordingly, the mask for determining the Neumann neighborhood: $01010101_{(2)} = 55_{(16)}$.

The above algorithm for one iteration of the CA is implemented by the following fragment of the source code on Apendix 3 is presented.

CONCLUSION

The developed technology made it possible to create a model of the interaction of many active cells with different active states in the ACA space. In this case, the state of the active cell is determined by the state of the cells of the neighborhood. Based on this approach, the dynamics of the appearance (birth) and disappearance (dying) of active cells during their interaction is described. The studies for different forms of interaction made it possible to describe a dynamic model of living organisms at a primitive level. This ACA structure allows us to describe the various forms of behavior of distributed intelligent systems. The proposed technique complements the theoretical positions in the field of cellular automata. Software implementation makes it possible to set various behavioral modes of active cells, as well as form forecasts for overpopulation processes.

REFERENCES

Adamatzky, A. (2010). *Game of life Cellular automata*. Springer-Verlag London. doi:10.1007/978-1-84996-217-9

Adamatzky, A. (2018). Cellular automata. Springer Science + Business Media LLC.

Gardner, M. (1970). The fantastic combinations of John Conway's new solitaire game "Life". *Scientific American, N4*, 120–123. doi:10.1038cientificamerican1070-120

Komosinnski, M., & Adamatzky, A. (2009). *Artifical Life Models in Software* (2nd ed.). Springer. doi:10.1007/978-1-84882-285-6

Snitkovsky S. (1972). Once again about the "evolution". *Science and Life, 8*, 141-144.

APPENDIX

The cellular automaton simulation program with a variable number of active cells is written in MS Visual C # 2008 using the working environment of the NET Framework 2.0. The program has a convenient and easy to understand interface. The main form of the program consists of a tabbed area and a cellular automaton field. Between them is a counter of the current number of active cells. The main code fragments describing the CA and its functions are presented in this Appendix.

The fragment of the source code responsible for creating the CA, its initial initialization and visualization:

```
int kx = Convert.ToInt32(textBox1.Text);
int ky = Convert.ToInt32(textBox2.Text);
bool size_changed = false;
if (KAx !=  kx)
{
    size_changed = true;
    KAx = kx;
}
if (KAy !=  ky)
{
    size_changed = true;
    KAy = ky;
}
int cell_size = 22;
radioButton8.Enabled = true;
KA = new int [KAx,KAy];
if (size_changed)
{
    panel1.Visible = false;
    panel1.Controls.Clear();
    panel1.Width = KAx * cell_size + 4;
    panel1.Height = KAy * cell_size + 4;
    step = 0;
    pnl = new Label[KAx, KAy];
    Array.Clear(pnl, 0, pnl.Length);
    if (AC_count > 0)
    {
        Array.Clear(arrayAC, 0, AC_count);
        AC_count = 0;
    }
```

```
    for (int y = 0; y < KAy; y++)
    {
        for (int x = 0; x < KAx; x++)
        {
            int index = y * KAx + x;
            pnl[x, y] = new System.Windows.Forms.Label();
            pnl[x, y].Location = new System.Drawing.Point(1 + x
* (cell_size), 1 + y * (cell_size));
            //pnl[x, y].Name = "label_pnl_" + (index + 1000).
ToString();
            pnl[x, y].Name = "lblpnl_" + x.ToString() + "_" +
y.ToString();
            toolTip1.SetToolTip(pnl[x, y], "cell [" +
x.ToString() + ", " + y.ToString() + "]");
            pnl[x, y].AutoSize = false;
            pnl[x, y].Size = new System.Drawing.Size(cell_size,
cell_size);
            pnl[x, y].TextAlign = ContentAlignment.
MiddleCenter;
            pnl[x, y].MinimumSize = new System.Drawing.
Size(cell_size, cell_size);
            pnl[x, y].Font = new Font("Lucida Console", 8);
            pnl[x, y].BorderStyle = BorderStyle.FixedSingle;
            pnl[x, y].KeyUp += new KeyEventHandler(tb_KeyDown);
            pnl[x, y].MouseClick += new MouseEventHandler(tb_
MouseClick);
            panel1.Controls.Add(pnl[x, y]);
        }
    }
}
else
{
    if (radioButton8.Checked)
    {
    }
}
Array.Clear(KA, 0, KA.Length);
for (int y = 0; y < KAy; y++)
{
    for (int x = 0; x < KAx; x++)
    {
        if (radioButton3.Checked)
        {
            KA[x, y] = Convert.ToInt16(rnd.NextDouble());
        }
        else if (radioButton2.Checked)
        {
            KA[x, y] = 0;
        }
```

```
        else if (radioButton1.Checked)
        {
            KA[x, y] = 1;
        }
        else if (radioButton8.Checked)
        {
            if (pnl[x, y].Text == "0")
            {
                KA[x, y] = 1;
            }
            else
            {
                KA[x, y] = 0;
            }
        }
        pnl[x, y].Text = KA[x, y].ToString();
        if (KA[x, y] == 0)
        {
            pnl[x, y].BackColor = color_0;
        }
        else
        {
            pnl[x, y].BackColor = color_1;
        }
    }
}
panel1.Visible = true;
panel1.Refresh();
```

The active cell corresponds to a structure consisting of its coordinates in the CA, mask displacement code and local function:

```
public struct AC
{
    public int X;
    public int Y;
    public int dvig_mask;
    public int LSF;
};
```

The algorithm for one iteration of the CA is implemented by the following fragment of the source code:

```
step = step + 1;
label12.Text = step.ToString();
bool deletion_present = false;
AC_count = arrayAC.Length;
```

```
do
{
    deletion_present = false;
    for (int i = 0; i < AC_count - 1; i++)
    {
        for (int q = i + 1; q < AC_count; q++)
        {
            if (((arrayAC[i].X == arrayAC[q].X) &&
(arrayAC[i].Y == arrayAC[q].Y)) ||
                (!isInside_mode && isSosedi(arrayAC[i].X,
arrayAC[i].Y, arrayAC[q].X, arrayAC[q].Y)))
            {
                if ((arrayAC[i].dvig_mask == arrayAC[q].dvig_
mask))
                {
                    arrayAC_decolored_index(q);
                    arrayAC_del_index(q);
                    arrayAC_decolored_index(i);
                    arrayAC_del_index(i);
                    deletion_present = true;
                    break;
                }
            }
            if (deletion_present) break;
        }
    }
} while (deletion_present);
AC_count = arrayAC.Length;
for (int i = 0; i < AC_count - 1; i++)
{
    for (int q = i + 1; q < AC_count; q++)
    {
        if (isInside_mode && (arrayAC[i].X == arrayAC[q].X) &&
(arrayAC[i].Y == arrayAC[q].Y) ||
            !isInside_mode && isSosedi(arrayAC[i].X,
arrayAC[i].Y, arrayAC[q].X, arrayAC[q].Y))
        {
            int pi;
            int pq;
            int newX = -1;
            int newY = -1;
            int dvig_mask_new = 0x00;
            bool must_add_new_AC = true;
            if (!isInside_mode)
            {
                pi = get_perehod_value(i);
                pq = get_perehod_value(q);
                if (pi > pq)
                {
```

```
                        newX = arrayAC[i].X;
                        newY = arrayAC[i].Y;
                }
                else if (pi < pq)
                {
                        newX = arrayAC[q].X;
                        newY = arrayAC[q].Y;
                }
                else
                {
                        must_add_new_AC = false;
                }
        }
        else
        {
            newX = arrayAC[i].X;
            newY = arrayAC[i].Y;
        }
        if (!isFullInteractionAC_mode && must_add_new_AC)
        {
            if (get_count_AC_in_this(newX, newY) >= 2)
            {
                must_add_new_AC = false;
            }
        }
        if (must_add_new_AC)
        {
            if (isNeumann)
            {
                if ((arrayAC[i].dvig_mask == arrayAC[q].
dvig_mask))
                {
                    must_add_new_AC = false;
                }
                else if ((arrayAC[i].dvig_mask ^
arrayAC[q].dvig_mask) == 0x55)
                {
                    must_add_new_AC = false;
                }
                else
                {
                    dvig_mask_new = (arrayAC[i].dvig_mask ^
arrayAC[q].dvig_mask) & pokr_mask;
                }
            }
            else
            {
                int x_bits = (arrayAC[i].dvig_mask &
arrayAC[q].dvig_mask) & pokr_mask;
```

```
                    int x_bits_1 = Convert.ToString(x_bits,
2).PadLeft(8, '0').Replace("0", "").Length;
                    if (x_bits_1 == 1)
                    {
                        x_bits = (arrayAC[i].dvig_mask |
arrayAC[q].dvig_mask) ^ pokr_mask;
                        dvig_mask_new = x_bits & pokr_mask;
                    }
                    else if (x_bits_1 == 2)
                    {
                        if (isFullInteractionAC_mode)
                        {
                            x_bits = (arrayAC[i].dvig_mask ^
arrayAC[q].dvig_mask);
                            int x_bit0 = Convert.
ToString(((arrayAC[i].dvig_mask | arrayAC[q].dvig_mask) &
pokr_mask), 2).PadLeft(8, '0').LastIndexOf("0");
                            x_bit0 = 7 - x_bit0;
                            dvig_mask_new = (x_bits | (0x1 <<
x_bit0)) & pokr_mask;
                        }
                        else
                        {
                            must_add_new_AC = false;
                        }
                    }
                    else
                    {
                        must_add_new_AC = false;
                    }
                }
                if (!isFullInteractionAC_mode && must_add_new_
AC)
                {
                    if (must_add_new_AC)
                    {
                        for (int a = 0; a < arrayAC.Length;
a++)
                        {
                            if ((arrayAC[a].dvig_mask == dvig_
mask_new) && (arrayAC[a].X == newX) && (arrayAC[a].Y == newY))
                            {
                                must_add_new_AC = false;
                                break;
                            }
                        }
                    }
                }
            }
```

```
            if (must_add_new_AC)
            {
                Array.Resize(ref arrayAC, arrayAC.Length + 1);
                int idnew = arrayAC.Length - 1;
                arrayAC[idnew].X = newX;
                arrayAC[idnew].Y = newY;
                arrayAC[idnew].LSF = (arrayAC[i].LSF +
arrayAC[q].LSF + 1) % 3;
                arrayAC[idnew].dvig_mask = dvig_mask_new;
            }
        }
    }
}
if (isInside_mode) AC_count = arrayAC.Length;
int[,] AC_new_values;
int soss;
AC_new_values = new int[AC_count, 3];
for (int i = 0; i < AC_count; i++)
{
    int dvig = get_perehod_value(i);
    soss = get_sosedi(arrayAC[i].X, arrayAC[i].Y);
    bool lsf = Convert.ToBoolean(KA[arrayAC[i].X,
arrayAC[i].Y]);
    for (int b = 0; b < 8; b++)
    {
        if (((pokr_mask >> b) & 0x01) == 0x01)
        {
            bool sbit = Convert.ToBoolean(((soss) >> b) &
0x01);
            if (arrayAC[i].LSF == 0) //XOR
            {
                lsf = lsf ^ sbit;
            }
            else if (arrayAC[i].LSF == 1) //AND
            {
                lsf = lsf & sbit;
            }
            else if (arrayAC[i].LSF == 2) //OR
            {
                lsf = lsf | sbit;
            }
        }
    }
    AC_new_values[i, 0] = arrayAC[i].X;
    AC_new_values[i, 1] = arrayAC[i].Y;
    AC_new_values[i, 2] = Convert.ToInt32(lsf);
    pnl[arrayAC[i].X, arrayAC[i].Y].Text = AC_new_values[i,
2].ToString();
    arrayAC_decolored_index(i);
```

```
    int[] perehod = get_sosed_koordinates(arrayAC[i].X,
arrayAC[i].Y, dvig);
    arrayAC[i].X = perehod[0];
    arrayAC[i].Y = perehod[1];
}
if (ordinary_cell_function > 0)
{
    int[,] KA_ = new int[KAx, KAy];
    for (int y = 0; y < KAy; y++)
    {
        for (int x = 0; x < KAx; x++)
        {
            bool isACxy = false;
            for (int i = 0; i < AC_count; i++)
            {
                if ((AC_new_values[i, 0] == x) && (AC_new_
values[i, 1] == y))
                {
                    isACxy = true;
                    break;
                }
            }
            if (!isACxy)
            {
                soss = get_sosedi(x, y);
                bool ocf = Convert.ToBoolean(KA[x, y]);
                for (int b = 0; b < 8; b++)
                {
                    if (((pokr_mask >> b) & 0x01) == 0x01)
                    {
                        bool sbit = Convert.ToBoolean(((soss)
>> b) & 0x01);
                        if (ordinary_cell_function == 1) //XOR
                        {
                            ocf = ocf ^ sbit;
                        }
                        else if (ordinary_cell_function == 2)
//AND
                        {
                            ocf = ocf & sbit;
                        }
                        else if (ordinary_cell_function == 3)
//OR
                        {
                            ocf = ocf | sbit;
                        }
                    }
                }
                KA_[x, y] = Convert.ToInt16(ocf);
```

```
                  if (KA[x, y] != KA_[x, y])
                  {
                      pnl[x, y].Text = KA_[x,y].ToString();
                      if (KA_[x, y] == 0)
                      {
                          pnl[x, y].BackColor = color_0;
                      }
                      else
                      {
                          pnl[x, y].BackColor = color_1;
                      }
                  }
              }
          }
      }
   Array.Copy(KA_, KA, KA_.Length);
   Array.Clear(KA_, 0, KA_.Length);
}
for (int i = 0; i < AC_count; i++)
{
    KA[AC_new_values[i, 0], AC_new_values[i, 1]] = AC_new_
values[i, 2];
    pnl[AC_new_values[i, 0], AC_new_values[i, 1]].Text = AC_
new_values[i, 2].ToString();
}
if ((checkBox_perStep.Checked) && (step % Convert.
ToInt32(comboBox1.Text) == 0))
{
    for (int i = 0; i < AC_count; i++)
    {
        int col_shift_bit = 1;
        if (isNeumann)
        {
            col_shift_bit = 2;
        }
        arrayAC[i].dvig_mask = arrayAC[i].dvig_mask << col_
shift_bit | arrayAC[i].dvig_mask >> (8 - col_shift_bit);
        arrayAC[i].dvig_mask != arrayAC[i].dvig_mask & pokr_
mask;
    }
}
AC_count = arrayAC.Length;
for (int i = 0; i < AC_count; i++)
{
    arrayAC_colored_index(i);
}
for (int i = 0; i < AC_count; i++)
{
    arrayAC_colored_index(i, false);
```

```
}
label5.Text = AC_count.ToString();
if (checkBox_epidemic.Checked)
{
    if (AC_count > (KAx * KAy) / 2)
    {
        if (timer1.Enabled)
        {
            timer1.Stop();
            timer1.Enabled = false;
            button3.Text = "Start";
        }
        tabControl1.SelectedIndex = 2;
        MessageBox.Show("Attention! Epidemic! The number of
active cells has exceeded half of all CA cells! Inspection
soon!")))");
    }
}
```

Another feature of the modeling program is the unique layout of controls (controls), which allows to fit all the necessary functionality of the controls and display tools on one main form. Below is the corresponding listing of Form1.Designer.cs:

```
namespace CA_MultiAct
{
    partial class Form1
    {
        /// <summary>
        /// Required designer variable.
        /// </summary>
        private System.ComponentModel.IContainer components =
null;
        /// <summary>
        /// Clean up any resources being used.
        /// </summary>
        /// <param name="disposing">true if managed resources
should be disposed; otherwise, false.</param>
        protected override void Dispose(bool disposing)
        {
            if (disposing && (components != null))
            {
                components.Dispose();
            }
            base.Dispose(disposing);
        }
        #region Windows Form Designer generated code
```

```
        /// <summary>
        /// Required method for Designer support - do not
modify
        /// the contents of this method with the code editor.
        /// </summary>
        private void InitializeComponent()
        {
            this.components = new System.ComponentModel.
Container();
            System.ComponentModel.ComponentResourceManager
resources = new System.ComponentModel.ComponentResourceManager(
typeof(Form1));
            this.button1 = new System.Windows.Forms.Button();
            this.panel1 = new System.Windows.Forms.Panel();
            this.tabControl1 = new System.Windows.Forms.
TabControl();
            this.tabPage1 = new System.Windows.Forms.TabPage();
            this.groupBox4 = new System.Windows.Forms.
GroupBox();
            this.radioButton10 = new System.Windows.Forms.
RadioButton();
            this.radioButton9 = new System.Windows.Forms.
RadioButton();
            this.button5 = new System.Windows.Forms.Button();
            this.groupBox2 = new System.Windows.Forms.
GroupBox();
            this.radioButton5 = new System.Windows.Forms.
RadioButton();
            this.radioButton4 = new System.Windows.Forms.
RadioButton();
            this.groupBox1 = new System.Windows.Forms.
GroupBox();
            this.radioButton8 = new System.Windows.Forms.
RadioButton();
            this.radioButton3 = new System.Windows.Forms.
RadioButton();
            this.radioButton2 = new System.Windows.Forms.
RadioButton();
            this.radioButton1 = new System.Windows.Forms.
RadioButton();
            this.label2 = new System.Windows.Forms.Label();
            this.label1 = new System.Windows.Forms.Label();
            this.textBox2 = new System.Windows.Forms.TextBox();
            this.textBox1 = new System.Windows.Forms.TextBox();
            this.tabPage2 = new System.Windows.Forms.TabPage();
            this.label10 = new System.Windows.Forms.Label();
            this.label8 = new System.Windows.Forms.Label();
            this.label7 = new System.Windows.Forms.Label();
            this.listBox1 = new System.Windows.Forms.ListBox();
```

```
            this.label6 = new System.Windows.Forms.Label();
            this.label3 = new System.Windows.Forms.Label();
            this.comboBox2 = new System.Windows.Forms.
ComboBox();
            this.button4 = new System.Windows.Forms.Button();
            this.button2 = new System.Windows.Forms.Button();
            this.checkBox_dvig5 = new System.Windows.Forms.
CheckBox();
            this.checkBox_dvig4 = new System.Windows.Forms.
CheckBox();
            this.checkBox_dvig3 = new System.Windows.Forms.
CheckBox();
            this.checkBox_dvig6 = new System.Windows.Forms.
CheckBox();
            this._dvig = new System.Windows.Forms.CheckBox();
            this.checkBox_dvig7 = new System.Windows.Forms.
CheckBox();
            this.checkBox_dvig0 = new System.Windows.Forms.
CheckBox();
            this.checkBox_dvig2 = new System.Windows.Forms.
CheckBox();
            this.checkBox_dvig1 = new System.Windows.Forms.
CheckBox();
            this.tabPage3 = new System.Windows.Forms.TabPage();
            this.checkBox2 = new System.Windows.Forms.
CheckBox();
            this.checkBox_epidemic = new System.Windows.Forms.
CheckBox();
            this.checkBox1 = new System.Windows.Forms.
CheckBox();
            this.comboBox1 = new System.Windows.Forms.
ComboBox();
            this.label9 = new System.Windows.Forms.Label();
            this.checkBox_perStep = new System.Windows.Forms.
CheckBox();
            this.button3 = new System.Windows.Forms.Button();
            this.groupBox3 = new System.Windows.Forms.
GroupBox();
            this.textBox3 = new System.Windows.Forms.TextBox();
            this.radioButton7 = new System.Windows.Forms.
RadioButton();
            this.radioButton6 = new System.Windows.Forms.
RadioButton();
            this.label5 = new System.Windows.Forms.Label();
            this.label4 = new System.Windows.Forms.Label();
            this.menuStrip1 = new System.Windows.Forms.
MenuStrip();
            this.toolTip1 = new System.Windows.Forms.
ToolTip(this.components);
```

```
            this.timer1 = new System.Windows.Forms.Timer(this.
components);
            this.comboBox3 = new System.Windows.Forms.
ComboBox();
            this.label11 = new System.Windows.Forms.Label();
            this.label12 = new System.Windows.Forms.Label();
            this.label13 = new System.Windows.Forms.Label();
            this.tabControl1.SuspendLayout();
            this.tabPage1.SuspendLayout();
            this.groupBox4.SuspendLayout();
            this.groupBox2.SuspendLayout();
            this.groupBox1.SuspendLayout();
            this.tabPage2.SuspendLayout();
            this.tabPage3.SuspendLayout();
            this.groupBox3.SuspendLayout();
            this.SuspendLayout();
            //
            // button1
            //
            this.button1.Location = new System.Drawing.
Point(337, 41);
            this.button1.Name = "button1";
            this.button1.Size = new System.Drawing.Size(128,
31);
            this.button1.TabIndex = 0;
            this.button1.Text = "Build new CA";
            this.button1.UseVisualStyleBackColor = true;
            this.button1.Click += new System.EventHandler(this.
button1_Click);
            //
            // panel1
            //
            this.panel1.BackColor = System.Drawing.
SystemColors.ControlDarkDark;
            this.panel1.BorderStyle = System.Windows.Forms.
BorderStyle.FixedSingle;
            this.panel1.Location = new System.Drawing.Point(4,
110);
            this.panel1.Name = "panel1";
            this.panel1.Size = new System.Drawing.Size(70, 73);
            this.panel1.TabIndex = 1;
            //
            // tabControl1
            //
            this.tabControl1.Controls.Add(this.tabPage1);
            this.tabControl1.Controls.Add(this.tabPage2);
            this.tabControl1.Controls.Add(this.tabPage3);
            this.tabControl1.Location = new System.Drawing.
Point(0, 0);
```

```
            this.tabControl1.Name = "tabControl1";
            this.tabControl1.SelectedIndex = 0;
            this.tabControl1.Size = new System.Drawing.
Size(484, 104);
            this.tabControl1.TabIndex = 2;
            this.tabControl1.Selected += new System.Windows.
Forms.TabControlEventHandler(this.tabControl1_Selected);
            this.tabControl1.MouseDoubleClick += new System.
Windows.Forms.MouseEventHandler(this.tabControl1_
MouseDoubleClick);
            //
            // tabPage1
            //
            this.tabPage1.BackColor = System.Drawing.Color.
Transparent;
            this.tabPage1.Controls.Add(this.label11);
            this.tabPage1.Controls.Add(this.button5);
            this.tabPage1.Controls.Add(this.comboBox3);
            this.tabPage1.Controls.Add(this.groupBox4);
            this.tabPage1.Controls.Add(this.groupBox2);
            this.tabPage1.Controls.Add(this.groupBox1);
            this.tabPage1.Controls.Add(this.label2);
            this.tabPage1.Controls.Add(this.label1);
            this.tabPage1.Controls.Add(this.textBox2);
            this.tabPage1.Controls.Add(this.textBox1);
            this.tabPage1.Controls.Add(this.button1);
            this.tabPage1.Location = new System.Drawing.
Point(4, 22);
            this.tabPage1.Name = "tabPage1";
            this.tabPage1.Padding = new System.Windows.Forms.
Padding(3);
            this.tabPage1.Size = new System.Drawing.Size(476,
78);
            this.tabPage1.TabIndex = 0;
            this.tabPage1.Text = "CA design";
            this.tabPage1.UseVisualStyleBackColor = true;
            //
            // groupBox4
            //
            this.groupBox4.Controls.Add(this.radioButton10);
            this.groupBox4.Controls.Add(this.radioButton9);
            this.groupBox4.Location = new System.Drawing.
Point(167, 39);
            this.groupBox4.Name = "groupBox4";
            this.groupBox4.Size = new System.Drawing.Size(161,
33);
            this.groupBox4.TabIndex = 6;
            this.groupBox4.TabStop = false;
            this.groupBox4.Text = "AC interaction mode";
```

```
            this.toolTip1.SetToolTip(this.groupBox4, "active
cell interaction mode");
            this.groupBox4.Enter += new System.
EventHandler(this.groupBox4_Enter);
            //
            // radioButton10
            //
            this.radioButton10.AutoSize = true;
            this.radioButton10.Location = new System.Drawing.
Point(89, 10);
            this.radioButton10.Name = "radioButton10";
            this.radioButton10.Size = new System.Drawing.
Size(56, 17);
            this.radioButton10.TabIndex = 0;
            this.radioButton10.Text = "beside";
            this.radioButton10.UseVisualStyleBackColor = true;
            this.radioButton10.CheckedChanged += new System.
EventHandler(this.radioButton10_CheckedChanged);
            //
            // radioButton9
            //
            this.radioButton9.AutoSize = true;
            this.radioButton9.Checked = true;
            this.radioButton9.Location = new System.Drawing.
Point(6, 10);
            this.radioButton9.Name = "radioButton9";
            this.radioButton9.Size = new System.Drawing.
Size(52, 17);
            this.radioButton9.TabIndex = 0;
            this.radioButton9.TabStop = true;
            this.radioButton9.Text = "inside";
            this.radioButton9.UseVisualStyleBackColor = true;
            this.radioButton9.CheckedChanged += new System.
EventHandler(this.radioButton9_CheckedChanged);
            //
            // button5
            //
            this.button5.Location = new System.Drawing.
Point(92, 10);
            this.button5.Name = "button5";
            this.button5.Size = new System.Drawing.Size(22,
30);
            this.button5.TabIndex = 4;
            this.button5.Text = "button5";
            this.button5.UseVisualStyleBackColor = true;
            this.button5.Visible = false;
            this.button5.Click += new System.EventHandler(this.
button5_Click);
            //
```

```
            // groupBox2
            //
            this.groupBox2.Controls.Add(this.radioButton5);
            this.groupBox2.Controls.Add(this.radioButton4);
            this.groupBox2.Location = new System.Drawing.
Point(12, 38);
            this.groupBox2.Name = "groupBox2";
            this.groupBox2.Size = new System.Drawing.Size(144,
35);
            this.groupBox2.TabIndex = 5;
            this.groupBox2.TabStop = false;
            this.groupBox2.Text = "Neighborhood";
            this.toolTip1.SetToolTip(this.groupBox2, "тип
покрытия");
            //
            // radioButton5
            //
            this.radioButton5.AutoSize = true;
            this.radioButton5.Location = new System.Drawing.
Point(86, 12);
            this.radioButton5.Name = "radioButton5";
            this.radioButton5.Size = new System.Drawing.
Size(55, 17);
            this.radioButton5.TabIndex = 1;
            this.radioButton5.Text = "Moore";
            this.radioButton5.UseVisualStyleBackColor = true;
            this.radioButton5.CheckedChanged += new System.
EventHandler(this.radioButton5_CheckedChanged);
            //
            // radioButton4
            //
            this.radioButton4.AutoSize = true;
            this.radioButton4.Checked = true;
            this.radioButton4.Location = new System.Drawing.
Point(12, 12);
            this.radioButton4.Name = "radioButton4";
            this.radioButton4.Size = new System.Drawing.
Size(71, 17);
            this.radioButton4.TabIndex = 0;
            this.radioButton4.TabStop = true;
            this.radioButton4.Text = "Neumann";
            this.radioButton4.UseVisualStyleBackColor = true;
            this.radioButton4.CheckedChanged += new System.
EventHandler(this.radioButton4_CheckedChanged);
            //
            // groupBox1
            //
            this.groupBox1.Controls.Add(this.radioButton8);
            this.groupBox1.Controls.Add(this.radioButton3);
```

```
            this.groupBox1.Controls.Add(this.radioButton2);
            this.groupBox1.Controls.Add(this.radioButton1);
            this.groupBox1.Location = new System.Drawing.
Point(262, 2);
            this.groupBox1.Name = "groupBox1";
            this.groupBox1.Size = new System.Drawing.Size(208,
35);
            this.groupBox1.TabIndex = 4;
            this.groupBox1.TabStop = false;
            this.groupBox1.Text = "fill:";
            //
            // radioButton8
            //
            this.radioButton8.AutoSize = true;
            this.radioButton8.Location = new System.Drawing.
Point(171, 13);
            this.radioButton8.Name = "radioButton8";
            this.radioButton8.Size = new System.Drawing.
Size(32, 17);
            this.radioButton8.TabIndex = 1;
            this.radioButton8.TabStop = true;
            this.radioButton8.Text = "~";
            this.toolTip1.SetToolTip(this.radioButton8, "
current state inversion ");
            this.radioButton8.UseVisualStyleBackColor = true;
            //
            // radioButton3
            //
            this.radioButton3.AutoSize = true;
            this.radioButton3.Checked = true;
            this.radioButton3.Location = new System.Drawing.
Point(6, 13);
            this.radioButton3.Name = "radioButton3";
            this.radioButton3.Size = new System.Drawing.
Size(49, 17);
            this.radioButton3.TabIndex = 0;
            this.radioButton3.TabStop = true;
            this.radioButton3.Text = "RND";
            this.toolTip1.SetToolTip(this.radioButton3, "
randomly fill");
            this.radioButton3.UseVisualStyleBackColor = true;
            //
            // radioButton2
            //
            this.radioButton2.AutoSize = true;
            this.radioButton2.Location = new System.Drawing.
Point(66, 13);
            this.radioButton2.Name = "radioButton2";
            this.radioButton2.Size = new System.Drawing.
```

```
Size(41, 17);
            this.radioButton2.TabIndex = 0;
            this.radioButton2.Text = "\"0\"";
            this.toolTip1.SetToolTip(this.radioButton2, " fill
with zeros");
            this.radioButton2.UseVisualStyleBackColor = true;
            //
            // radioButton1
            //
            this.radioButton1.AutoSize = true;
            this.radioButton1.Location = new System.Drawing.
Point(117, 13);
            this.radioButton1.Name = "radioButton1";
            this.radioButton1.Size = new System.Drawing.
Size(41, 17);
            this.radioButton1.TabIndex = 0;
            this.radioButton1.Text = "\"1\"";
            this.toolTip1.SetToolTip(this.radioButton1, " fill
in units");
            this.radioButton1.UseVisualStyleBackColor = true;
            //
            // label2
            //
            this.label2.AutoSize = true;
            this.label2.Location = new System.Drawing.Point(6,
2);
            this.label2.Name = "label2";
            this.label2.Size = new System.Drawing.Size(28, 13);
            this.label2.TabIndex = 3;
            this.label2.Text = "size:";
            //
            // label1
            //
            this.label1.AutoSize = true;
            this.label1.Location = new System.Drawing.Point(42,
17);
            this.label1.Name = "label1";
            this.label1.Size = new System.Drawing.Size(12, 13);
            this.label1.TabIndex = 2;
            this.label1.Text = "x";
            //
            // textBox2
            //
            this.textBox2.Location = new System.Drawing.
Point(56, 14);
            this.textBox2.Name = "textBox2";
            this.textBox2.Size = new System.Drawing.Size(30,
20);
            this.textBox2.TabIndex = 1;
```

```
          this.textBox2.Text = "16";
          this.textBox2.TextAlign = System.Windows.Forms.
HorizontalAlignment.Center;
          this.textBox2.TextChanged += new System.
EventHandler(this.textBox2_TextChanged);
          //
          // textBox1
          //
          this.textBox1.Location = new System.Drawing.
Point(9, 14);
          this.textBox1.Name = "textBox1";
          this.textBox1.Size = new System.Drawing.Size(30,
20);
          this.textBox1.TabIndex = 1;
          this.textBox1.Text = "16";
          this.textBox1.TextAlign = System.Windows.Forms.
HorizontalAlignment.Center;
          this.textBox1.TextChanged += new System.
EventHandler(this.textBox1_TextChanged);
          //
          // tabPage2
          //
          this.tabPage2.BackColor = System.Drawing.Color.
Transparent;
          this.tabPage2.Controls.Add(this.label10);
          this.tabPage2.Controls.Add(this.label8);
          this.tabPage2.Controls.Add(this.label7);
          this.tabPage2.Controls.Add(this.listBox1);
          this.tabPage2.Controls.Add(this.label6);
          this.tabPage2.Controls.Add(this.label3);
          this.tabPage2.Controls.Add(this.comboBox2);
          this.tabPage2.Controls.Add(this.button4);
          this.tabPage2.Controls.Add(this.button2);
          this.tabPage2.Controls.Add(this.checkBox_dvig5);
          this.tabPage2.Controls.Add(this.checkBox_dvig4);
          this.tabPage2.Controls.Add(this.checkBox_dvig3);
          this.tabPage2.Controls.Add(this.checkBox_dvig6);
          this.tabPage2.Controls.Add(this._dvig);
          this.tabPage2.Controls.Add(this.checkBox_dvig7);
          this.tabPage2.Controls.Add(this.checkBox_dvig0);
          this.tabPage2.Controls.Add(this.checkBox_dvig2);
          this.tabPage2.Controls.Add(this.checkBox_dvig1);
          this.tabPage2.Location = new System.Drawing.
Point(4, 22);
          this.tabPage2.Name = "tabPage2";
          this.tabPage2.Padding = new System.Windows.Forms.
Padding(3);
          this.tabPage2.Size = new System.Drawing.Size(476,
78);
```

287

```
            this.tabPage2.TabIndex = 1;
            this.tabPage2.Text = "Active cells";
            this.toolTip1.SetToolTip(this.tabPage2, " double
click on the title - build a complete list of AC at the current
time" +
                    "and");
            this.tabPage2.UseVisualStyleBackColor = true;
            this.tabPage2.Leave += new System.
EventHandler(this.tabPage2_Leave);
            //
            // label10
            //
            this.label10.AutoSize = true;
            this.label10.Location = new System.Drawing.
Point(216, 62);
            this.label10.Name = "label10";
            this.label10.Size = new System.Drawing.Size(13,
13);
            this.label10.TabIndex = 13;
            this.label10.Text = "0";
            this.toolTip1.SetToolTip(this.label10, "number of
active cells in this list");
            //
            // label8
            //
            this.label8.AutoSize = true;
            this.label8.Location = new System.Drawing.Point(8,
11);
            this.label8.Name = "label8";
            this.label8.Size = new System.Drawing.Size(24, 13);
            this.label8.TabIndex = 12;
            this.label8.Text = "Cell";
            //
            // label7
            //
            this.label7.AutoSize = true;
            this.label7.Location = new System.Drawing.Point(2,
28);
            this.label7.Name = "label7";
            this.label7.Size = new System.Drawing.Size(30, 13);
            this.label7.TabIndex = 12;
            this.label7.Text = "[#,#]";
            this.toolTip1.SetToolTip(this.label7, "coordinates
of the current Ctrl-clicked cell");
            //
            // listBox1
            //
            this.listBox1.FormattingEnabled = true;
            this.listBox1.HorizontalScrollbar = true;
```

```
            this.listBox1.Location = new System.Drawing.
Point(46, 5);
            this.listBox1.Name = "listBox1";
            this.listBox1.Size = new System.Drawing.Size(178,
69);
            this.listBox1.TabIndex = 11;
            this.listBox1.SelectedIndexChanged += new System.
EventHandler(this.listBox1_SelectedIndexChanged);
            //
            // label6
            //
            this.label6.AutoSize = true;
            this.label6.Location = new System.Drawing.
Point(230, 8);
            this.label6.Name = "label6";
            this.label6.Size = new System.Drawing.Size(62, 13);
            this.label6.TabIndex = 10;
            this.label6.Text = "code mask:";
            //
            // label3
            //
            this.label3.AutoSize = true;
            this.label3.Location = new System.Drawing.
Point(309, 19);
            this.label3.Name = "label3";
            this.label3.Size = new System.Drawing.Size(29, 13);
            this.label3.TabIndex = 9;
            this.label3.Text = "LSF:";
            //
            // comboBox2
            //
            this.comboBox2.DropDownStyle = System.Windows.
Forms.ComboBoxStyle.DropDownList;
            this.comboBox2.FormattingEnabled = true;
            this.comboBox2.Items.AddRange(new object[] {
            "XOR",
            "AND",
            "OR"});
            this.comboBox2.Location = new System.Drawing.
Point(290, 35);
            this.comboBox2.Name = "comboBox2";
            this.comboBox2.Size = new System.Drawing.Size(71,
21);
            this.comboBox2.TabIndex = 8;
            //
            // button4
            //
            this.button4.Location = new System.Drawing.
Point(367, 33);
```

```
            this.button4.Name = "button4";
            this.button4.Size = new System.Drawing.Size(100,
23);
            this.button4.TabIndex = 5;
            this.button4.Text = "Remove cell";
            this.button4.UseVisualStyleBackColor = true;
            this.button4.Click += new System.EventHandler(this.
button4_Click);
            //
            // button2
            //
            this.button2.Location = new System.Drawing.
Point(367, 8);
            this.button2.Name = "button2";
            this.button2.Size = new System.Drawing.Size(100,
23);
            this.button2.TabIndex = 5;
            this.button2.Text = "Add active cell";
            this.button2.UseVisualStyleBackColor = true;
            this.button2.Click += new System.EventHandler(this.
button2_Click);
            //
            // checkBox_dvig5
            //
            this.checkBox_dvig5.AutoSize = true;
            this.checkBox_dvig5.Location = new System.Drawing.
Point(235, 52);
            this.checkBox_dvig5.Name = "checkBox_dvig5";
            this.checkBox_dvig5.Size = new System.Drawing.
Size(15, 14);
            this.checkBox_dvig5.TabIndex = 4;
            this.checkBox_dvig5.UseVisualStyleBackColor = true;
            this.checkBox_dvig5.CheckedChanged += new System.
EventHandler(this.checkBox_dvig5_CheckedChanged);
            //
            // checkBox_dvig4
            //
            this.checkBox_dvig4.AutoSize = true;
            this.checkBox_dvig4.Location = new System.Drawing.
Point(250, 52);
            this.checkBox_dvig4.Name = "checkBox_dvig4";
            this.checkBox_dvig4.Size = new System.Drawing.
Size(15, 14);
            this.checkBox_dvig4.TabIndex = 4;
            this.checkBox_dvig4.UseVisualStyleBackColor = true;
            this.checkBox_dvig4.CheckedChanged += new System.
EventHandler(this.checkBox_dvig4_CheckedChanged);
            //
            // checkBox_dvig3
```

```
            //
            this.checkBox_dvig3.AutoSize = true;
            this.checkBox_dvig3.Location = new System.Drawing.
Point(265, 52);
            this.checkBox_dvig3.Name = "checkBox_dvig3";
            this.checkBox_dvig3.Size = new System.Drawing.
Size(15, 14);
            this.checkBox_dvig3.TabIndex = 4;
            this.checkBox_dvig3.UseVisualStyleBackColor = true;
            this.checkBox_dvig3.CheckedChanged += new System.
EventHandler(this.checkBox_dvig3_CheckedChanged);
            //
            // checkBox_dvig6
            //
            this.checkBox_dvig6.AutoSize = true;
            this.checkBox_dvig6.Location = new System.Drawing.
Point(235, 38);
            this.checkBox_dvig6.Name = "checkBox_dvig6";
            this.checkBox_dvig6.Size = new System.Drawing.
Size(15, 14);
            this.checkBox_dvig6.TabIndex = 3;
            this.checkBox_dvig6.UseVisualStyleBackColor = true;
            this.checkBox_dvig6.CheckedChanged += new System.
EventHandler(this.checkBox_dvig6_CheckedChanged);
            //
            // _dvig
            //
            this._dvig.AutoSize = true;
            this._dvig.Enabled = false;
            this._dvig.Location = new System.Drawing.Point(250,
38);
            this._dvig.Name = "_dvig";
            this._dvig.Size = new System.Drawing.Size(15, 14);
            this._dvig.TabIndex = 3;
            this._dvig.UseVisualStyleBackColor = true;
            //
            // checkBox_dvig7
            //
            this.checkBox_dvig7.AutoSize = true;
            this.checkBox_dvig7.Location = new System.Drawing.
Point(235, 24);
            this.checkBox_dvig7.Name = "checkBox_dvig7";
            this.checkBox_dvig7.Size = new System.Drawing.
Size(15, 14);
            this.checkBox_dvig7.TabIndex = 2;
            this.checkBox_dvig7.UseVisualStyleBackColor = true;
            this.checkBox_dvig7.CheckedChanged += new System.
EventHandler(this.checkBox_dvig7_CheckedChanged);
            //
```

```
            // checkBox_dvig0
            //
            this.checkBox_dvig0.AutoSize = true;
            this.checkBox_dvig0.Location = new System.Drawing.
Point(250, 24);
            this.checkBox_dvig0.Name = "checkBox_dvig0";
            this.checkBox_dvig0.Size = new System.Drawing.
Size(15, 14);
            this.checkBox_dvig0.TabIndex = 2;
            this.checkBox_dvig0.UseVisualStyleBackColor = true;
            this.checkBox_dvig0.CheckedChanged += new System.
EventHandler(this.checkBox_dvig0_CheckedChanged);
            //
            // checkBox_dvig2
            //
            this.checkBox_dvig2.AutoSize = true;
            this.checkBox_dvig2.Location = new System.Drawing.
Point(265, 38);
            this.checkBox_dvig2.Name = "checkBox_dvig2";
            this.checkBox_dvig2.Size = new System.Drawing.
Size(15, 14);
            this.checkBox_dvig2.TabIndex = 3;
            this.checkBox_dvig2.UseVisualStyleBackColor = true;
            this.checkBox_dvig2.CheckedChanged += new System.
EventHandler(this.checkBox_dvig2_CheckedChanged);
            //
            // checkBox_dvig1
            //
            this.checkBox_dvig1.AutoSize = true;
            this.checkBox_dvig1.Location = new System.Drawing.
Point(265, 24);
            this.checkBox_dvig1.Name = "checkBox_dvig1";
            this.checkBox_dvig1.Size = new System.Drawing.
Size(15, 14);
            this.checkBox_dvig1.TabIndex = 2;
            this.checkBox_dvig1.UseVisualStyleBackColor = true;
            this.checkBox_dvig1.CheckedChanged += new System.
EventHandler(this.checkBox_dvig1_CheckedChanged);
            //
            // tabPage3
            //
            this.tabPage3.BackColor = System.Drawing.Color.
Transparent;
            this.tabPage3.Controls.Add(this.checkBox2);
            this.tabPage3.Controls.Add(this.checkBox_epidemic);
            this.tabPage3.Controls.Add(this.checkBox1);
            this.tabPage3.Controls.Add(this.comboBox1);
            this.tabPage3.Controls.Add(this.label9);
            this.tabPage3.Controls.Add(this.checkBox_perStep);
```

```
            this.tabPage3.Controls.Add(this.button3);
            this.tabPage3.Controls.Add(this.groupBox3);
            this.tabPage3.Location = new System.Drawing.
Point(4, 22);
            this.tabPage3.Name = "tabPage3";
            this.tabPage3.Size = new System.Drawing.Size(476,
78);
            this.tabPage3.TabIndex = 2;
            this.tabPage3.Text = "CA run";
            this.tabPage3.UseVisualStyleBackColor = true;
            //
            // checkBox2
            //
            this.checkBox2.AutoSize = true;
            this.checkBox2.Location = new System.Drawing.
Point(180, 25);
            this.checkBox2.Name = "checkBox2";
            this.checkBox2.Size = new System.Drawing.Size(120,
17);
            this.checkBox2.TabIndex = 8;
            this.checkBox2.Text = "full interaction of AC";
            this.toolTip1.SetToolTip(this.checkBox2, "full
interaction of all ACs with no restrictions on the appearance
of the 3rd active function in the cell and interaction with the
2-3 mask according to Moore");
            this.checkBox2.UseVisualStyleBackColor = true;
            this.checkBox2.CheckedChanged += new System.
EventHandler(this.checkBox2_CheckedChanged);
            //
            // checkBox_epidemic
            //
            this.checkBox_epidemic.AutoSize = true;
            this.checkBox_epidemic.Checked = true;
            this.checkBox_epidemic.CheckState = System.Windows.
Forms.CheckState.Checked;
            this.checkBox_epidemic.Location = new System.
Drawing.Point(180, 44);
            this.checkBox_epidemic.Name = "checkBox_epidemic";
            this.checkBox_epidemic.Size = new System.Drawing.
Size(112, 17);
            this.checkBox_epidemic.TabIndex = 7;
            this.checkBox_epidemic.Text = "\'epidemic\'
warning";
            this.toolTip1.SetToolTip(this.checkBox_epidemic,
"Warning of a large number of ACs and automatic mode stop");
            this.checkBox_epidemic.UseVisualStyleBackColor =
true;
            //
            // checkBox1
```

```
              //
              this.checkBox1.AutoSize = true;
              this.checkBox1.Location = new System.Drawing.
Point(375, 56);
              this.checkBox1.Name = "checkBox1";
              this.checkBox1.Size = new System.Drawing.Size(92,
17);
              this.checkBox1.TabIndex = 6;
              this.checkBox1.Text = "hide CA panel";
              this.toolTip1.SetToolTip(this.checkBox1, "Hide the
panel with a cellular automata - removes the brakes with a
large number of AC");
              this.checkBox1.UseVisualStyleBackColor = true;
              this.checkBox1.CheckedChanged += new System.
EventHandler(this.checkBox1_CheckedChanged);
              //
              // comboBox1
              //
              this.comboBox1.FormattingEnabled = true;
              this.comboBox1.Items.AddRange(new object[] {
              "1",
              "2",
              "3",
              "4",
              "5",
              "6",
              "7",
              "8",
              "9",
              "10",
              "11",
              "12",
              "13",
              "14",
              "15"});
              this.comboBox1.Location = new System.Drawing.
Point(285, 3);
              this.comboBox1.Name = "comboBox1";
              this.comboBox1.Size = new System.Drawing.Size(35,
21);
              this.comboBox1.TabIndex = 5;
              this.comboBox1.Text = "3";
              //
              // label9
              //
              this.label9.AutoSize = true;
              this.label9.Location = new System.Drawing.
Point(323, 6);
              this.label9.Name = "label9";
```

```
            this.label9.Size = new System.Drawing.Size(32, 13);
            this.label9.TabIndex = 4;
            this.label9.Text = "steps";
            //
            // checkBox_perStep
            //
            this.checkBox_perStep.AutoSize = true;
            this.checkBox_perStep.Location = new System.
Drawing.Point(180, 5);
            this.checkBox_perStep.Name = "checkBox_perStep";
            this.checkBox_perStep.Size = new System.Drawing.
Size(110, 17);
            this.checkBox_perStep.TabIndex = 2;
            this.checkBox_perStep.Text = "rotate mask every";
            this.toolTip1.SetToolTip(this.checkBox_perStep,
"Cyclic circular shift of mask of the moving code AC every N
steps");
            this.checkBox_perStep.UseVisualStyleBackColor =
true;
            this.checkBox_perStep.CheckedChanged += new System.
EventHandler(this.checkBox_perStep_CheckedChanged);
            //
            // button3
            //
            this.button3.Location = new System.Drawing.
Point(365, 9);
            this.button3.Name = "button3";
            this.button3.Size = new System.Drawing.Size(102,
41);
            this.button3.TabIndex = 1;
            this.button3.Text = "Start";
            this.button3.UseVisualStyleBackColor = true;
            this.button3.Click += new System.EventHandler(this.
button3_Click);
            //
            // groupBox3
            //
            this.groupBox3.Controls.Add(this.textBox3);
            this.groupBox3.Controls.Add(this.radioButton7);
            this.groupBox3.Controls.Add(this.radioButton6);
            this.groupBox3.Location = new System.Drawing.
Point(8, 5);
            this.groupBox3.Name = "groupBox3";
            this.groupBox3.Size = new System.Drawing.Size(166,
60);
            this.groupBox3.TabIndex = 0;
            this.groupBox3.TabStop = false;
            this.groupBox3.Text = "run mode";
            //
```

```
        // textBox3
        //
        this.textBox3.Location = new System.Drawing.
Point(120, 36);
        this.textBox3.Name = "textBox3";
        this.textBox3.Size = new System.Drawing.Size(40,
20);
        this.textBox3.TabIndex = 2;
        this.textBox3.Text = "100";
        this.textBox3.TextAlign = System.Windows.Forms.
HorizontalAlignment.Center;
        this.toolTip1.SetToolTip(this.textBox3, "delay
between iterations in milliseconds");
        //
        // radioButton7
        //
        this.radioButton7.AutoSize = true;
        this.radioButton7.Location = new System.Drawing.
Point(6, 37);
        this.radioButton7.Name = "radioButton7";
        this.radioButton7.Size = new System.Drawing.
Size(118, 17);
        this.radioButton7.TabIndex = 1;
        this.radioButton7.TabStop = true;
        this.radioButton7.Text = "auto with delay, ms:";
        this.radioButton7.UseVisualStyleBackColor = true;
        //
        // radioButton6
        //
        this.radioButton6.AutoSize = true;
        this.radioButton6.Checked = true;
        this.radioButton6.Location = new System.Drawing.
Point(6, 19);
        this.radioButton6.Name = "radioButton6";
        this.radioButton6.Size = new System.Drawing.
Size(82, 17);
        this.radioButton6.TabIndex = 1;
        this.radioButton6.TabStop = true;
        this.radioButton6.Text = "step by step";
        this.radioButton6.UseVisualStyleBackColor = true;
        //
        // label5
        //
        this.label5.AutoSize = true;
        this.label5.Location = new System.Drawing.
Point(250, 97);
        this.label5.Name = "label5";
        this.label5.Size = new System.Drawing.Size(13, 13);
        this.label5.TabIndex = 7;
```

```
            this.label5.Text = "0";
            this.toolTip1.SetToolTip(this.label5, "Total number
of AC");
            //
            // label4
            //
            this.label4.AutoSize = true;
            this.label4.Location = new System.Drawing.
Point(124, 97);
            this.label4.Name = "label4";
            this.label4.Size = new System.Drawing.Size(104,
13);
            this.label4.TabIndex = 6;
            this.label4.Text = "Total AC / iterations:";
            //
            // menuStrip1
            //
            this.menuStrip1.Location = new System.Drawing.
Point(0, 0);
            this.menuStrip1.Name = "menuStrip1";
            this.menuStrip1.Size = new System.Drawing.Size(483,
24);
            this.menuStrip1.TabIndex = 3;
            this.menuStrip1.Text = "menuStrip1";
            //
            // timer1
            //
            this.timer1.Tick += new System.EventHandler(this.
timer1_Tick);
            //
            // comboBox3
            //
            this.comboBox3.DropDownStyle = System.Windows.
Forms.ComboBoxStyle.DropDownList;
            this.comboBox3.FormattingEnabled = true;
            this.comboBox3.Items.AddRange(new object[] {
            "no function",
            "XOR",
            "OR",
            "AND"});
            this.comboBox3.Location = new System.Drawing.
Point(137, 16);
            this.comboBox3.Name = "comboBox3";
            this.comboBox3.Size = new System.Drawing.Size(88,
21);
            this.comboBox3.TabIndex = 7;
            this.toolTip1.SetToolTip(this.comboBox3, "normal
(inactive) cell function");
            this.comboBox3.SelectedIndexChanged += new System.
```

```
EventHandler(this.comboBox3_SelectedIndexChanged);
            //
            // label11
            //
            this.label11.AutoSize = true;
            this.label11.Location = new System.Drawing.
Point(116, 3);
            this.label11.Name = "label11";
            this.label11.Size = new System.Drawing.Size(124,
13);
            this.label11.TabIndex = 8;
            this.label11.Text = "function of ordinary cells:";
            //
            // label12
            //
            this.label12.AutoSize = true;
            this.label12.Location = new System.Drawing.
Point(290, 97);
            this.label12.Name = "label12";
            this.label12.Size = new System.Drawing.Size(13,
13);
            this.label12.TabIndex = 7;
            this.label12.Text = "0";
            //
            // label13
            //
            this.label13.AutoSize = true;
            this.label13.Location = new System.Drawing.
Point(279, 97);
            this.label13.Name = "label13";
            this.label13.Size = new System.Drawing.Size(12,
13);
            this.label13.TabIndex = 7;
            this.label13.Text = "/";
            //
            // Form1
            //
            this.AutoScaleDimensions = new System.Drawing.
SizeF(6F, 13F);
            this.AutoScaleMode = System.Windows.Forms.
AutoScaleMode.Font;
            this.AutoSize = true;
            this.AutoSizeMode = System.Windows.Forms.
AutoSizeMode.GrowAndShrink;
            this.ClientSize = new System.Drawing.Size(483,
529);
            this.Controls.Add(this.label13);
            this.Controls.Add(this.label12);
            this.Controls.Add(this.label5);
```

```
            this.Controls.Add(this.label4);
            this.Controls.Add(this.tabControl1);
            this.Controls.Add(this.panel1);
            this.Controls.Add(this.menuStrip1);
            this.Icon = ((System.Drawing.Icon)(resources.
GetObject("$this.Icon")));
            this.MainMenuStrip = this.menuStrip1;
            this.Name = "Form1";
            this.Text = "CA_MultiAct";
            this.tabControl1.ResumeLayout(false);
            this.tabPage1.ResumeLayout(false);
            this.tabPage1.PerformLayout();
            this.groupBox4.ResumeLayout(false);
            this.groupBox4.PerformLayout();
            this.groupBox2.ResumeLayout(false);
            this.groupBox2.PerformLayout();
            this.groupBox1.ResumeLayout(false);
            this.groupBox1.PerformLayout();
            this.tabPage2.ResumeLayout(false);
            this.tabPage2.PerformLayout();
            this.tabPage3.ResumeLayout(false);
            this.tabPage3.PerformLayout();
            this.groupBox3.ResumeLayout(false);
            this.groupBox3.PerformLayout();
            this.ResumeLayout(false);
            this.PerformLayout();
        }
        #endregion
        private System.Windows.Forms.Button button1;
        private System.Windows.Forms.Panel panel1;
        private System.Windows.Forms.TabControl tabControl1;
        private System.Windows.Forms.TabPage tabPage1;
        private System.Windows.Forms.TabPage tabPage2;
        private System.Windows.Forms.MenuStrip menuStrip1;
        private System.Windows.Forms.ToolTip toolTip1;
        private System.Windows.Forms.TextBox textBox1;
        private System.Windows.Forms.TextBox textBox2;
        private System.Windows.Forms.GroupBox groupBox1;
        private System.Windows.Forms.RadioButton radioButton3;
        private System.Windows.Forms.RadioButton radioButton2;
        private System.Windows.Forms.RadioButton radioButton1;
        private System.Windows.Forms.Label label2;
        private System.Windows.Forms.Label label1;
        private System.Windows.Forms.CheckBox checkBox_dvig5;
        private System.Windows.Forms.CheckBox checkBox_dvig4;
        private System.Windows.Forms.CheckBox checkBox_dvig3;
        private System.Windows.Forms.CheckBox checkBox_dvig6;
        private System.Windows.Forms.CheckBox checkBox_dvig7;
        private System.Windows.Forms.CheckBox checkBox_dvig0;
```

```
       private System.Windows.Forms.CheckBox checkBox_dvig2;
       private System.Windows.Forms.CheckBox checkBox_dvig1;
       private System.Windows.Forms.CheckBox _dvig;
       private System.Windows.Forms.GroupBox groupBox2;
       private System.Windows.Forms.RadioButton radioButton5;
       private System.Windows.Forms.RadioButton radioButton4;
       private System.Windows.Forms.Button button2;
       private System.Windows.Forms.TabPage tabPage3;
       private System.Windows.Forms.GroupBox groupBox3;
       private System.Windows.Forms.RadioButton radioButton7;
       private System.Windows.Forms.RadioButton radioButton6;
       private System.Windows.Forms.Button button3;
       private System.Windows.Forms.Label label5;
       private System.Windows.Forms.Label label4;
       private System.Windows.Forms.Button button4;
       private System.Windows.Forms.Label label3;
       private System.Windows.Forms.ComboBox comboBox2;
       private System.Windows.Forms.Label label6;
       private System.Windows.Forms.ListBox listBox1;
       private System.Windows.Forms.Timer timer1;
       private System.Windows.Forms.TextBox textBox3;
       private System.Windows.Forms.Label label7;
       private System.Windows.Forms.Label label8;
       private System.Windows.Forms.Button button5;
       private System.Windows.Forms.CheckBox checkBox_perStep;
       private System.Windows.Forms.Label label9;
       private System.Windows.Forms.ComboBox comboBox1;
       private System.Windows.Forms.Label label10;
       private System.Windows.Forms.CheckBox checkBox1;
       private System.Windows.Forms.CheckBox checkBox_
epidemic;
       private System.Windows.Forms.RadioButton radioButton8;
       private System.Windows.Forms.GroupBox groupBox4;
       private System.Windows.Forms.RadioButton radioButton10;
       private System.Windows.Forms.RadioButton radioButton9;
       private System.Windows.Forms.CheckBox checkBox2;
       private System.Windows.Forms.ComboBox comboBox3;
       private System.Windows.Forms.Label label11;
       private System.Windows.Forms.Label label12;
       private System.Windows.Forms.Label label13;
    }
}
```

Chapter 9
Models and Paradigms of Cellular Automata With an Organized Set of Active Cells

ABSTRACT

The chapter presents the principles of functioning of asynchronous cellular automata with a group of cells united in a colony. The rules of the formation of colonies of active cells and methods to move them along the field of a cellular automaton are considered. Each formed colony of active cells has a main cell that controls the movement of the entire colony. If several colonies of identical cells meet and combine, then the main cell is selected according to the priority, which is evaluated by the state of the cells of their neighborhoods. Colonies with different active cells can interact, destroying each other. The methods of interaction of colonies with different active states are described. An example of colony formation for solving the problem of describing contour images is presented. The image is described by moving the colony through the cells belonging to the image contour and fixing the cell sectors of the colony, which include the cells of the contour at each time step.

INTRODUCTION

In the previous chapters, CA models were considered, which described the evolution of CA during the interaction of individual active cells. In this case, models of the formation of cell colonies with active states were not

DOI: 10.4018/978-1-7998-2649-1.ch009

considered. Not considered the interaction of groups of cells with the same and different active states. There is also a need to simulate the processes of formation of colonies and their decay, taking into account the behavior of various insects, animals and and people.

The chapter presents the principles of functioning of asynchronous cellular automata with a group of cells united in a colony. The rules of the formation of colonies of active cells and methods to move them along the field of a cellular automaton are considered. Each formed colony of active cells has a main cell that controls the movement of the entire colony. If several colonies of identical cells meet and combine, then the main cell is selected according to the priority, which is evaluated by the state of the cells of their neighborhoods. Colonies with different active cells can interact, destroying each other. The methods of interaction of colonies with different active states are described. An example of colony formation for solving the problem of describing contour images is presented. The image is described by moving the colony through the cells belonging to the image contour and fixing the cell sectors of the colony, which include the cells of the contour at each time step.

BASIC THESES AND DEFINITIONS

In previous chapters, cellular automata were considered in which states changed with the help of active cells. ACA behaviors are described in which each active cell functions independently of other active cells. However, ACA structures are possible in which homogeneous groups of active cells (colonies of active cells) function. Homogeneous groups of active cells are characterized by the fact that they perform basic logical operations in interaction with other active cells. These cells can be neighboring (cells of a given shape of the neighborhood) and can be cells of the neighborhood of neighboring cells.

Interacting active cells can perform the same logical function, but can perform LSF, which depends on the performed LSF of neighboring active cells. A group of active cells is cells that are not isolated from each other. Cells of a group can interact with each other through other active cells of the same group. Examples of such groups of cells on Figure 1 are presented.

Figure 1 shows the groups of active cells that are highlighted in black. Each active cell of such a group can transmit active and informational signals to any other active cell of this group. These examples show that all active cells that make up the active groups have the same properties (have the same

Figure 1. Examples of groups of active cells

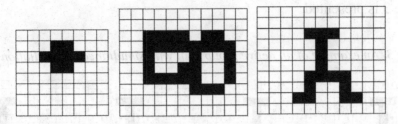

state). However, active groups are united not by basic informational states, but by states of activity. Examples of such groups on Figure 2 are presented.

In Figure 2, groups of different colors are formed from cells with different basic information states. Each group of active cells is highlighted in black. Thus, active groups are formed only by active cells.

Figure 2. Examples of groups of active cells with different informational states

1	0	1	0	1	0	0	1	0	0	1
0	1	0	0	0	0	0	0	1	0	0
1	0	0	1	0	0	1	0	0		0
0	1	1	1	1	0	0	1	1	0	1
1	0	1	1	0	0	0	1	0	0	1
0	1	0	1	0	0	1	1	1	1	1
1	0	1	1	0	0	0	0	1	1	1
0	1	1	1	0	0	0	1	1	0	0
1	0	1	0	1	0	0	1	0	0	1
0	1	0	0	0	0	0	0	1	0	0

1	0	1	0	1	0	0	1	0	0	1
0	1	0	0	0	0	0	0	1	0	0
0	0	0	0	0	1	0	0	0	1	1
0	1	0	0	0	1	0	0	1	0	1
0	0	1	1	0	0	0	1	0	1	1
0	1	0	0	0	1	0	0	1	0	0
1	0	0	0	0	1	0	1	0	0	0
1	1	1	0	0	1	0	1	0	0	0
0	0	0	0	1	0	1	0	0	0	1
0	1	1	0	0	0	1	0	0	0	1

However, in such groups (colonies) for the normal functioning of the groups, the active cells in them are divided into internal and extreme active cells. Edge active cells are called active cells, in which there are inactive cells in the vicinity, and internal active cells have only active cells in the neighborhood of the colony to which they belong.

The direction of movement of the group is determined by the edge cells of the group. In this case, all cells of the active group must determine the direction of movement. The movements of the cells of groups can be unidirectional and multi-directional. With unidirectional movement, all active cells in the group determine the same direction of movement. This can happen in one time cycle with one control signal, or it can happen in several time steps. An example of movement by all active cells during simultaneous installation on Figure 3 is presented.

Figure 3. An example of an active group shift with simultaneous installation

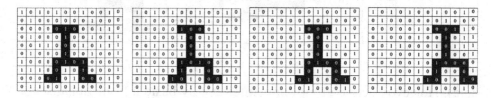

In this example, the shift of the active group is carried out rigidly and independently of the states of neighboring cells. If the shift of the cells of the group takes into account the states of neighboring cells, then the shift begins with the edge active cells of the group (Figure 4).

Figure 4. An example of a shift of the active cells of a group by one cell to the right, taking into account the states of neighboring cells

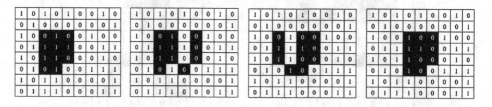

In the second embodiment, the extreme cells can detach from the entire group, since they are shifted to cells that do not belong to the active group. Here, cells can transmit an active signal only to neighboring inactive cells, which will set them to an active state.

The question arises. How cells in the appropriate active state are set? If external control is used, then the shift direction for cells of active groups is set by an external control signal, which violates the principle of cell colony self-organization.

Important is the independent choice of the direction of movement of the active cells of the entire active group. This enables the full autonomous functioning of the ACA. There is also the possibility of modeling the processes of interaction of colonies of active cells and track their evolution.

THE CHOICE OF THE DIRECTION OF MOTION

The choice of direction of movement of the colony is indicated by many different factors. Basically, edge colony cells take part in the choice of the direction of movement of the colony of active cells. These cells analyze the state of the cell environment outside the colony and choose the "favorable" direction of movement. They transmit the corresponding active signals to all cells of the colony, which establish the selected directions of movement in them. However, different extreme cells can choose different directions of movement of the colony, then the shape of the colony will stretch in different directions (Figure 5).

Figure 5. An example of choosing the direction of movement of different extreme cells of the colony

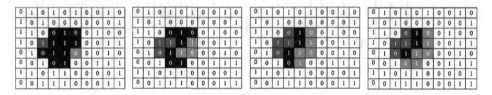

Green color shows cells that have a direction of movement to the left, and red color shows cells that have a direction of movement to the right. The color of the colony cells, which cannot determine the direction of movement, is shown in yellow. To propagate the active signal, the von Neumann neighborhood was used.

In the example shown in Figure 5, it can be seen that as a result, colony cells remain, which are difficult to choose the direction of movement. Therefore, it is necessary for such undecided cells to set a function that would make it possible to determine the direction of movement of the active cell. As such conditions, the following condition can be selected.

1. By the number of active cells in the neighborhood with the selected direction of shift.
2. By the number of active neighborhood cells that are in a logical "1" state (or logical "0" state) and have the same direction of movement.

For the first mode in the presented example, one cell remains with an indefinite direction of movement (Figure 6).

Figure 6. Result of using the first mode

0	1	0	1	0	1	0	0	1	0
1	0	1	0	0	0	0	0	0	1
1	1	0	0	1	0	0	1	0	0
1	0	1	1	1	1	0	0	1	1
0	1	0	1	1	0	0	0	1	0
0	0	1	0	1	0	0	1	1	1
1	1	0	1	1	0	0	0	0	1
0	0	1	1	1	0	0	0	1	1

The situation may also not change and the use of the second mode. The way out of this difficult situation is to use the eigenstate of the indefinite active cell. If the number of cells with different active states is the same, then the indefinite active cell takes the direction of the shift, which is determined by the inactive neighboring cell as a result of the shift of its active cell. If all the options described above are also the same, then the cell bifurcates and moves in two directions.

However, if it is necessary to move all active cells of the colony in only one direction, all edge active cells must make the same decision. To do this, all cells of the colony must be connected to each other and exchange information. First of all, all neighboring inactive cells of the colony cells are analyzed. These cells have a different underlying information state, which affects the edge cells of the colony.

One of the options for movement is movement towards the edge boundary inactive cell, which has a logical "1" state and is an edge single cell when going around the colony clockwise. It is important to choose the initial cell bypassing the colony since the colony can have any shape. The simplest is the arbitrary choice of the initial extreme active cell, from which begins the counting of boundary cells in a logical "1" state. Figure 7 shows an example of finding the direction of movement of a colony.

Figure 7. An example of the search direction of motion

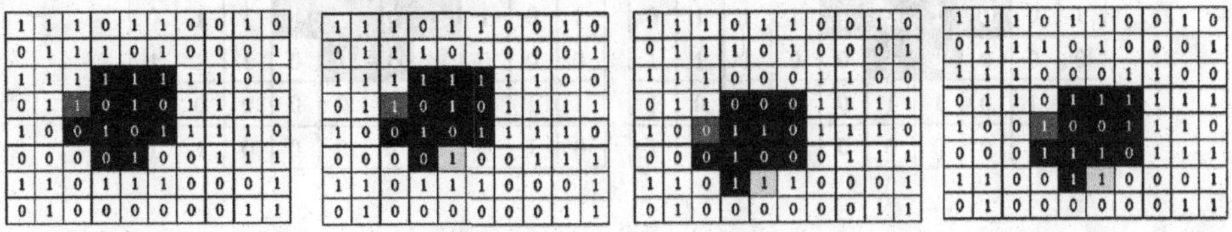

This figure shows the movement of the active signal in the extreme cells of the colony clockwise and counterclockwise. It is obvious that the edge inactive cell, bordering the colony cells and in the logical "1" state, is easier to determine by transmitting the active signal counterclockwise to the edge cells of the colony. The edge cell of the colony is determined that has a logical "1" state. In this case, there is no need to clockwise traverse the extreme cells of the colony. In the example shown in Figure 9.7, the colony's downward direction is determined.

After the colony cells shift one cell down, the next shift direction is determined, which corresponds to the right shift direction (Figure 7). To perform LSF (XOR function) by active colony cells and determine the direction of colony shift, the von Neumann neighborhood is used. In fact, the actual colony shift time step includes the time taken to determine the shift direction.

If you do not use the initial reference cell to bypass the edge cells of the colony, then you can implement several options for searching for the direction of movement. One of these options is to find favorable soil for the colony. The edge cells of the colony analyze neighboring cells that do not belong to the cells of the colony. For example, favorable soil is determined by the number of units in the neighborhood of the edge cells of the colony. In the direction where the number of cells having a logical "1" state is the largest shift of the colony at the next time step. Figure 8 shows an example of selecting the direction of movement of colony cells.

According to Figure 8, the direction of movement of the colony in the next time step is the direction to the right, since the edge right cells in the neighborhood have the most inactive cells that are in the logical "1" state. If situations arise when the same largest number of inactive single cells is present in several directions, the colony does not move in any of the shift

Figure 8. An example of independent choice of the direction of movement of the colony

1	1	1	0	1	1	0	0	1	0
0	1	0	1	0	1	1	0	0	1
1	1	1	1	1	1	1	1	0	0
0	1	1	0	1	0	1	1	1	1
1	0	0	1	0	1	1	1	1	0
0	0	0	0	1	0	0	1	1	1
1	1	0	1	1	0	0	0	0	1
0	1	0	0	0	0	0	0	1	1

1	1	1	0	1	1	0	0	1	0
0	1	0	1	0	1	1	0	0	1
1	1	1	0	0	0	1	1	0	0
0	1	1	0	0	0	1	1	1	1
1	0	1	1	0	0	1	1	1	0
0	0	0	1	0	0	0	1	1	1
1	1	0	1	1	0	0	0	0	1
0	1	0	0	0	0	0	0	1	1

directions. The shift begins when the state of inactive ACA cells changes. However, there is a need for a certain organization of active cells, which allows all active cells to determine the direction of the shift. At the moment, such an organization requires the search for optimal solutions and has still not found the right solution.

Another option is to use the main active cell of the colony, which would indicate the direction of movement for the active cells of the entire colony. The main active cell founded a colony or a cell that selected by "voting" all cells in the colony. If the founder of the colony is not difficult to determine, the task of organizing a "vote" has not yet been solved.

Assume that the main cell (the founder of the colony), which is the oldest active cell, is selected. The main cell analyzes neighborhood cells and determines the direction of movement of active neighborhood cells by the states of the first two neighborhood cells. The analysis of the first two cells of the neighborhood is performed for the von Neumann neighborhood. For Moore's neighborhood, an analysis of the states of the first three cells of the neighborhood is used.

Figure 9 shows an example of choosing the direction of movement based on the main senior cell of the colony.

Figure 9. An example of choosing the direction of movement based on the main active cell as the eldest founder of a colony of active cells

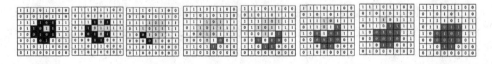

Figure 9 shows the process of transmitting to the colony cells an active direction signal. It can be seen that the individual edge cells (highlighted in red) have already shifted, and at this time there are active colony cells that have not yet received a directional signal. Therefore, Figure 9 shows how the movement of the entire colony is carried out with small gaps and the very shape of the colony stretches towards the direction of movement. After moving all the active cells of the colony, the shape of the colony becomes the same. Figure 9 describes the aperiodic process of the cellular automata. Therefore, the number of aperiodic steps, which are determined by the shape of the colony and the number of active cells organizing it stretch one time step of moving the colony to one ACA cell.

Using synchronization can be done in several modes:

- Synchronization only to move all cells of the colony;
- Synchronization for colony cells and for the shift of all colony cells.

In the first mode, the length of the period of formation of the clock signals is determined by the propagation time of the signal forming the direction of movement in all active cells of the colony. This mode limits the performance of the ACA. In the second mode, any functioning of each active and inactive cell is clocked.

The time of movement of a colony of active cells depends on the direction of movement and on the shape of the colony itself. The more colony cells are located on one straight line with the direction of movement, the slower the whole colony moves. A colony can contain a large number of active cells and move in one direction faster than in other directions of movement.

COLONY FORMATION

A colony of active cells can be set initially, if necessary to complete the task. However, there are tasks in which colonies can be created during the evolution of ACA. The previous chapter examined the process of colony formation from active cells with different active states. Moreover, the colony is formed from newly formed active cells and from existing active cells that have joined. The described colonies are unstable formations, so they constantly change their shape and have a variable number of active cells. In these colonies, active cells constantly disappear or new ones appear. This approach does not make it possible to achieve a stable colony shape.

Stable forms of the colony can be formed from active cells with identical active states. In this case, the condition for zeroing active cells with identical active states is not used when they interact as neighboring cells of the neighborhood. Thus, zeroing of identical active cells can occur only when they coincide in one active cell. In the case of the formation of a colony, such a situation cannot arise.

The following situation is used to form a colony. Old cells are present in the ACA field and new active cells appear. These cells have the same active states and move along the ACA field with a given LTF. If active cells with different active states meet and interact, then they form a new active cell with a new active state. If neighboring cells become active cells with the same active states, then they unite and form a colony of identical active cells. The process of colony formation on Figure 10 is shown.

Figure 10. Example of the colony formation process

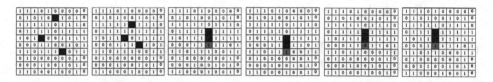

In Figure 10, three active cells with equal active states are present in the ACA field. At the third time step, three active cells become adjacent in two. They belong to the cells of the neighborhood of each of them. The middle cell is adjacent to two active cells. At this point, in the third time step, these three cells form a colony. The main cell (highlighted in red) in this colony becomes a cell that has more cells in the neighborhood that are in a logical "1" state. The own state of the active cell is also taken into account. Further, at each time step, a colony of three cells moves in the directions determined by the main cell. This cell remains the main one at all further time steps.

Colonies with the same active cells can merge with each other. Moreover, there is a need to determine the main cell from the two main cells of the colony. The most affordable options for implementing the definition of the main active cell is:

- By the largest number of cells that form a colony;
- By the largest number of cells that are in a logical "1" state and belong to the neighborhood of the main cell of the colony;

- Selection of the main cell of the newly formed colony, which is not one of the previous main cells.

Figure 11 shows examples of colony merger according to the three options considered.

Figure 11. Examples of unite of colonies in three ways

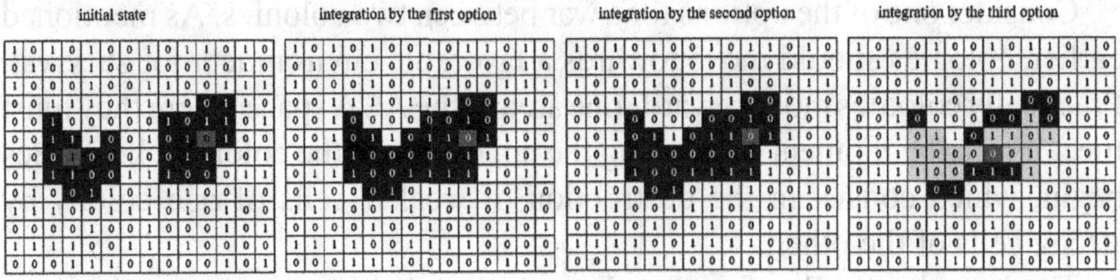

In each case, after the merger of the colonies, there remains one main cell (highlighted in red), which determines the direction of movement. In the third option, the cells through which the active signal passed. In the cell of the meeting of two active signals, a new main cell is determined. If there are several such cells, then one of the cells is selected that has more cells in the neighborhood that have a logical state of "1". If such cells (having a logical "1" state) are equal in the neighborhood of both cells, then the colonies do not merge into one colony. For the first option, the colonies also do not merge if they consist of an equal number of active cells. Similarly, for the second option, the colonies do not merge if the main cells contain the same number of single cells in the neighborhood.

There are many more options for combining colonies. It all depends on the initial values, as well as on the chosen forms of neighborhoods, LSF and LTF.

COLONY INTERACTION

The previous section examined the processes of colony unification. However, processes of the decay of colonies into individual cells or into smaller colonies are possible. Destruction of colonies or war between colonies may also occur.

The number of cells forming the colony can determine the decay process of the colonies. If the number of cells exceeds a certain value, then the decay of the colony begins. This decay can be defined in different ways. For example, groups of cells to which the control signal arrives later than all other cells can be determined. For separated groups of cells, it is important to select a new main cell. Options for choosing new main cells are discussed in the previous sections of the chapter.

If several colonies of cells with different active states are found, a conflict situation may arise that leads to the destruction of the cells of the colony. Also, hostilities can occur between colonies of active cells with the same active states.

Consider one of the options for a war between two colonies. As mentioned earlier, colonies should move towards a favorable environment for survival. Practically colonies move towards one area of the ACA space where they must meet. From the moment of meeting the edge cells of the colonies, hostilities begin, which consist in the destruction of weaker cells of one colony by strong cells of the other.

The boundary cells of each colony during their interaction analyze the state of the neighborhood cells and determine the number of neighborhood cells in a logical state of "1". Cells that have a larger number of single cells in a neighborhood survive, and neighboring cells belonging to another colony and having a smaller number of single cells in a neighborhood "die" (they become active or go into the active state of a strong cell in another colony, that is, they are captured). After the colonies touch, the colonies can move in a different direction. In this case, the initial forms of both colonies will be changed. An example of colony interaction on Figure 12 is shown.

Figure 12. An example of the interaction of colonies with different active cells

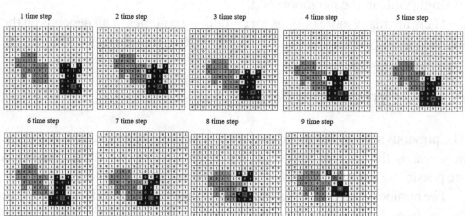

At the second time step (Figure 12), the two colonies begin to interact with each other. At this time step in each colony, one cell "dies". In the following time steps, the colonies continue to interact. Their structures change and at the eighth time step, the right colony is divided into three colonies, in each of which one main cell appears, which indicate their further direction of movement. The colony can disappear when all cells of the colony disappear or the main cell of the colony dies. Other colony interaction options are also possible. These options are implemented by ACA developers.

DESCRIPTION OF IMAGES USING CELL COLONIES

Cell colonies allow to describe contour images, as well as track various motion paths in real time. One option for using cell colonies is to describe the image by moving the cell colony of the selected shape along the path. For this purpose, a group of active cells is used, which moves along the contour with fixation of the breakage cells and cells belonging to the contour spaces. A graphical explanation of the method on Figure 13 is shown.

Figure 13. Example of changing and moving the target (cell colony) along the contour of the image

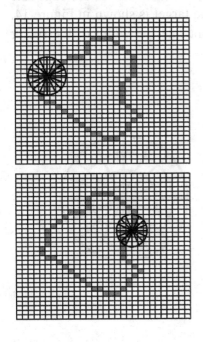
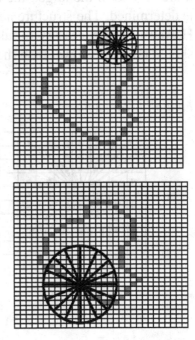

The method is as follows.

1. The image of the figure is projected onto the CA.
2. A target structure (cell colony shape) is formed that takes into account the number of its sectors and its radius.
3. Choosing the starting point of the edge and aligning it with the center of the target.
4. Search for sectors that are imposed on the neighboring cells.
5. Determination of the outermost edge cell of the target that fall into a specific sector.
6. Changing the size of the target (increasing or decreasing the radius) to coincide the edge of the target sector with the next breaking point of the contour.
7. The moves the target to the defined next contour point.
8. Search for the next breakpoint of the outline.

The process of finding the breakage cells continues until the center of the target is again compatible with the initial cell of the transfer of the active state.

The structure of the target is characterized by the number of sectors that indicate the corresponding angle of the edge line. The more sectors in the target, the more accurate the direction of the edge line is determined. The more sectors in the target, the more accurate the direction of the edge line is determined. The structure and shape of the target is shown in Figure 14. However, it is necessary to take into account that the width of the sector should not be narrower than the cell CA.

Figure 14. Target forms

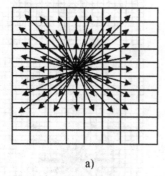

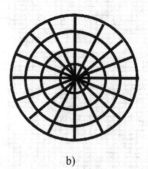

a)

b)

If the target consists of cells (Figure 14, a), the number of directions directly depends on the size (radius) of the colony. This significantly limits the possibilities of such a target.

If the target is made on the photodetector elements (Figure 14, b), then the accuracy and error of coding is increased. Each sector is made of a set of discrete photodetector sectors that capture the presence of cell affiliation to the image.

The selection of the initial cell of the edges is made by propagating the active signal across the KA field. When the active signal reaches the nearest cell of the image path (this is the initial cell), the process of moving the target on the edge line begins.

The target generates the matching cells of the edge cells with the corresponding direction. That is, a ones code of the direction of X_i is formed. In the process of moving the target is the addition of ones codes of cell to the general code that is formed during the movement.

An example of such encoding by time clock on Table 1 is given.

Table 1. Encoding by time-cycles

Time Cycles	Code	Comment
0	-------------	
1	X_1	
2	$X_1 X_1$	
3	$X_1 X_1 X_1$	
4	$X_1 X_1 X_1 X_4$	The point of breaking
5	$X_1 X_1 X_1 X_4 X_4$	
6	$X_1 X_1 X_1 X_4 X_4 X_4$	
. . .		
N	$X_1 X_1 X_1 X_4 X_4 X_4 \ldots\ldots\ldots X_7 X_7$	End of edge bypass

Changing of the size of the target occurs when the number of cells in one direction exceeds or less than the length of the radius of the target. In this case, the target may not change, but with its optimal value move along the path.

When determining the next cell of the edge fracture, the target is moved by the center of the cells belonging to the edge to the specified cell of the

Figure 15. Algorithm of cell functioning

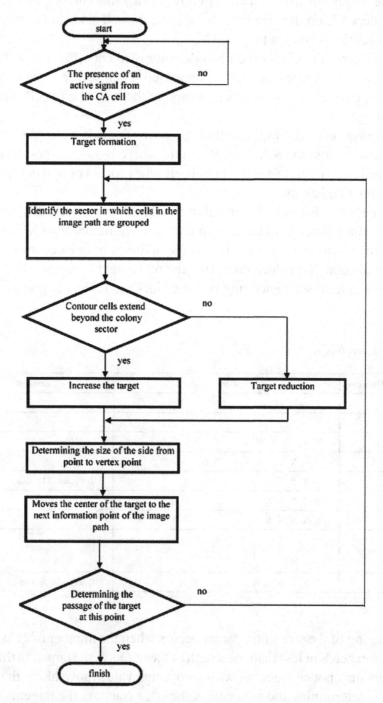

fracture. The number of cells in the same direction is taken into account during movement. The next break cell determines the new direction, as well as the number of cells and the subsequent break cell.

Thus, the code is generated according to the algorithm presented in Table 1. But such a code is unprofitable and therefore it is better to build code that is formed in the following form.

$$X_1 K_1 X_2 K_2 \ldots\ldots\ldots\ldots X_N K_N$$

In this form, X_i means direction, and K_i is the number of consecutive cells of the edge of the image with the directions of signal transmission of the image of X_i.

The software implementation of this method has difficulties in terms of virtual implementation of the target itself. That is, it would be preferable to implement the method at the hardware level.

The algorithm of cell functioning on Figure 15 is shown.

All ACA cells are waiting for an active signal to be received from the first CA edge information cell of the image contour, which became the original in the edge analysis. The structure of the target centered in the cell is formed on the basis of the ACA cell, which is the first to receive an active signal from the cell of the edge of image.

The formed target is divided into sectors of the cell placement directions. That is, a sector is defined by a group of cells that have a common logical - operational relationship in the group. Such a relation determines the size of the target and the direction.

When, by increasing or decreasing the target, the edge cell enters the adjacent sector from the specified one, then a signal is generated to move the center of the target to the specified cell. This process continues until a complete bypass of the edges of target.

Dividing a target into sectors of direction makes it possible to eliminate aliasing in recognition, which increases the accuracy of recognition of images invariant to rotation and scale.

Since the image of the object may be located freely unpredictably in the CA field, the initial cell (break point) of the edge is initially searched. This image was initially processed.

First of all, it eliminates noise and unnecessary image objects. The image is binarized and its edges removed. When the edges of the object image are selected, the active signal is propagated by the contour cells. This distribution is performed to find the initial cell of the edge that forms the top of the figure.

Figure 16. Second-order neighborhood encoding

X_{24}	X_9	X_{10}	X_{11}	X_{12}
X_{23}	X_8	X_1	X_2	X_{13}
X_{22}	X_7	X_0	X_3	X_{14}
X_{21}	X_6	X_5	X_4	X_{15}
X_{20}	X_{19}	X_{18}	X_{17}	X_{16}

Each cell of the edge is located in a place where the edge changes sharply. Such vertices are distinguished by the following coding (Figure 16).

The logical expression by which the cells encoding vertices are selected is as follows

Figure 17. Images with selected vertices

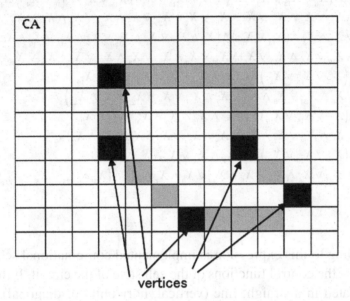

Figure 18. An example of the propagation of the active signal by the cells of the background

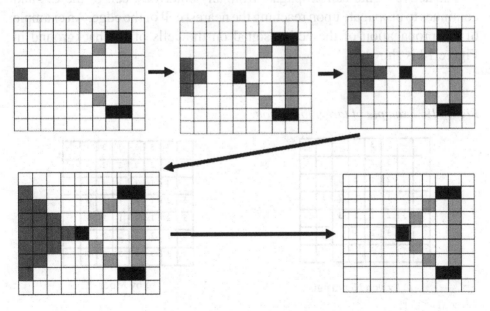

$$Q_{versh} = X_0 \wedge X_1 \wedge X_{10} \left(X_2 \wedge X_{11} \vee X_2 \wedge X_{12} \vee X_2 \wedge X_{13} \vee X_3 \wedge X_{13} \vee X_3 \wedge X_{14} \vee X_3 \wedge X_{15} \vee \right.$$
$$\vee X_4 \wedge X_{15} \vee X_4 \wedge X_{16} \vee X_4 \wedge X_{17} \vee X_6 \wedge X_{19} \vee X_6 \wedge X_{20} \vee X_6 \wedge X_{21} \vee$$
$$\vee X_7 \wedge X_{21} \vee X_7 \wedge X_{22} \vee X_7 \wedge X_{23} \vee X_8 \wedge X_{23} \vee X_8 \wedge X_{24} \vee X_8 \wedge X_9 \left. \right) \vee X_0 \wedge X_2$$
$$\wedge \left(X_{11} \vee X_{12} \right) \wedge \left(X_3 \wedge X_{13} \vee X_3 \wedge X_{14} \vee X_3 \wedge X_{15} \vee X_4 \wedge X_{15} \vee X_4 \wedge X_{16} \vee X_4 \wedge X_{17} \vee X_5 \right.$$
$$\wedge X_{17} \vee X_5 \wedge X_{18} \vee X_5 \wedge X_{19} \vee X_7 \wedge X_{21} \vee X_7 \wedge X_{22} \vee X_7 \wedge X_{23} \vee X_8 \wedge X_{23} \vee X_8 \wedge X_{24}$$
$$\vee X_8 \wedge X_9 \left. \right) \vee X_0 \wedge X_3 \wedge \left(X_{13} \vee X_{14} \right) \wedge \left(X_4 \wedge X_{15} \vee X_4 \wedge X_{16} \vee X_4 \wedge X_{17} \vee X_5 \wedge X_{17} \vee X_5 \right.$$
$$\wedge X_{18} \vee X_5 \wedge X_{19} \vee X_6 \wedge X_{19} \vee X_6 \wedge X_{20} \vee X_6 \wedge X_{21} \vee X_8 \wedge X_{23} \vee X_8 \wedge X_{24} \vee X_8 \wedge X_9 \left. \right) \vee$$
$$\vee X_0 \wedge X_4 \wedge \left(X_{15} \vee X_{16} \vee X_{17} \right) \wedge \left(X_5 \wedge X_{17} \vee X_5 \wedge X_{18} \vee X_5 \wedge X_{19} \vee \right.$$
$$\vee X_6 \wedge X_{19} \vee X_6 \wedge X_{20} \vee X_6 \wedge X_{21} \vee X_7 \wedge X_{21} \vee X_7 \wedge X_{22} \vee X_7 \wedge X_{23} \left. \right) \vee$$
$$\vee X_0 \wedge X_5 \wedge \left(X_{17} \vee X_{18} \vee X_{19} \right) \wedge \left(X_6 \wedge X_{19} \vee X_6 \wedge X_{20} \vee X_6 \wedge X_{21} \vee \right.$$
$$\vee X_7 \wedge X_{21} \vee X_7 \wedge X_{22} \vee X_7 \wedge X_{23} \vee X_8 \wedge X_{23} \vee X_8 \wedge X_{24} \vee X_8 \wedge X_9 \left. \right) \vee$$
$$\vee X_0 \wedge X_6 \wedge \left(X_{19} \vee X_{20} \vee X_{21} \right) \wedge \left(X_7 \wedge X_{21} \vee X_7 \wedge X_{22} \vee X_7 \right.$$
$$\wedge X_{23} \vee X_8 \wedge X_{23} \vee X_8 \wedge X_{24} \vee X_8 \wedge X_9 \left. \right) \vee X_0 \wedge X_7 \wedge X_{21} \wedge \left(X_8 \wedge X_{23} \vee X_8 \wedge X_{24} \vee X_8 \wedge X_9 \right)$$

According to this expression is implemented combinational scheme that implements the control functions of the cell state of the circuit. If the control cell is located in a straight line (vertical, horizontal or diagonal), it is not selected as a break point. An example of image with selected vertices on Figure 17 is shown.

The active signal can propagate from any outermost cell of the CA and continues to propagate upon reaching the nearest cell of the edge. An example of the propagation of the ective signal on the cells of the background in Figure 18 is shown.

Figure 19. Example of target formation

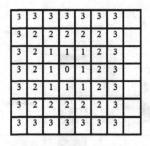

a) by von Neumann   b) by Moor

Upon reaching the active signal of the nearest selected cell of the edge at one of the outputs of this cell, a signal is generated, which is fed to the inputs of cells belonging to the background and prohibits the propagation of the active signal. From now on, the propagation of the active signal can only take place in the cells of edge of the image.

The form of increase of such a target depends on the organization of the neighborhood for the propagation of the active signal. An example of the

Figure 20. An example of determining basic attributes

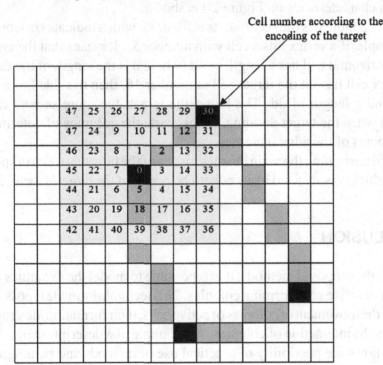

formation of a target by neighborhood of von Neumann and Moore on Figure 19 is shown.

In Figure 19 numbers denote imaginary cycles in which cells that attach to a target in a given cycle are selected. The target shapes are modified accordingly and search for the closest selected cell of the edge in different ways. The target is formed until one of the outermost cells of the target reaches one of the selected edge cells (vertices).

When this situation occurs, at the output of the selected cell, which received the signal of the target formation, a signal is generated, which is sent to all cells of the CA and blocks the process of formation of the target. From this point on, the process of forming the target stops. The determination of the distance between the central target cell (the selected vertex) and the vertex cell determined by the formation of the target begins.

The distance between the two cells is directly proportional to the number of imaginary cycles for which the target is formed. The slope or orientation of the side between the two cells is determined by the target cell that coincided with the vertex cell. An example of target formation and determination of the main characteristics on Figure 20 is shown.

Each cell indicates the sector of its affiliation, which indicates orientation. For example, if a vertex hits a cell with number 33, it means that the contour side is horizontal and has a length of 4d (where d is the length of the cell). If the vertex cell has hit the target cell at number 16, then the side has a slope of 45^0 and a length of 3d. The larger the target, the more sectors can be obtained when the target dividing. In this case, the accuracy of determining the attributes of the edge increases.

After determining the main characteristics of the target moves to a specific vertex, which was selected by its edge. The center of the target is located in it.

CONCLUSION

Based on the proposed methods, it was possible to model the dynamics of the behavior of active cells forming colonies. This technology made it possible to describe the movement of colonies of active cells, their formation, destruction, as well as the interaction of colonies. An example of a description of contour images shows the possibility of practical use of methods and paradigms for solving many problems. For modeling distributed intelligent systems for various purposes, this methodology occupies a special place.

About the Authors

Stepan Mykolayovych Bilan was born on September 15, 1962 in the Kazatin city, Vinnitsia region, Ukraine. He studied at the Vinnitsa Polytechnic Institute from 1979 to 1984. In 1984 he graduated from Vinnitsa Polytechnic with honors and an engineering diploma specializing in Electronic computing machines. From 1986 to 1989 he studied at the graduate school of the Vinnitsa Polytechnic Institute focusing on Computers, complexes and networks. In 1990 he defended his thesis by specialties: 05.13.13 - computer systems, complexes and networks; 05.13.05 - Elements and devices of computer facilities and control systems. He worked at Vinnytsia State Technical University (now the Vinnytsia National Technical University) from 1991 to 2003. In 1998 he was awarded the academic title of assistant professor of "Computer Science" at Vinnytsia National Technical University. From 2003 to the present time, he works at the State University of Infrastructure and Technology (Kiev, Ukraine).

Mykola Mykolayovych Bilan was born on June 27, 1961 in Kazatin, Vinnytsia region, Ukraine. In 1983 he graduated with honors from the Vinnitsa Polytechnic Institute and has a specialty Automation and Remote Control. Until 1994, he worked as a production master in the final assembly shop at the Tashkent Aviation Production Association. V.P. Chkalov. In 1989 he was the best master of the Association. From 1994 to 2015, he worked at the Radio and Television Center (Moldova Republic of) as an engineer and the head of the dispatch service. Currently, he work as an informatics and physics teacher in a secondary school in the village of Mayak, Moldova Republic of. He has many inventions, publications and patents in the field of cryptography, steganography and computer technology.

Ruslan Leonidovich Motornyuk was born on June 20, 1976 in the village of Ivankovtsy, Kazatinsky District, Vinnitsa Region, Ukrainian SSR. In 2001 he graduated from the magistracy of Vinnitsa State Technical University and received a master's degree in computer systems and networks. In 2013 he defended his thesis for the degree of candidate of technical sciences in the specialty "Computer systems and components". Currently he works as a leading engineer in the Production Unit "Kiev Department" branch of the Main Information and Computing Center of the JSC "Ukrzaliznytsya."

Index

Ensure Quality Research is Introduced to the Academic Community

Become an IGI Global Reviewer for Authored Book Projects

The overall success of an authored book project is dependent on quality and timely reviews.

In this competitive age of scholarly publishing, constructive and timely feedback significantly expedites the turnaround time of manuscripts from submission to acceptance, allowing the publication and discovery of forward-thinking research at a much more expeditious rate. Several IGI Global authored book projects are currently seeking highly-qualified experts in the field to fill vacancies on their respective editorial review boards:

Applications and Inquiries may be sent to:
development@igi-global.com

Applicants must have a doctorate (or an equivalent degree) as well as publishing and reviewing experience. Reviewers are asked to complete the open-ended evaluation questions with as much detail as possible in a timely, collegial, and constructive manner. All reviewers' tenures run for one-year terms on the editorial review boards and are expected to complete at least three reviews per term. Upon successful completion of this term, reviewers can be considered for an additional term.

If you have a colleague that may be interested in this opportunity, we encourage you to share this information with them.

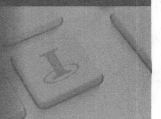

Printed in the United States
By Bookmasters